The Magic of Magnesium

Looks closely at magnesium's role as an important element for the maintenance of health, and investigates its value in combating osteoporosis, alleviating the symptoms of PMS and contributing towards the healthy functioning of the heart, arteries and kidneys.

The Magic of Magnesium

The element which
plays an important part in combating
osteoporosis and in alleviating the
symptoms of PMS

by

Dr Eric Trimmer

Thorsons
An Imprint of HarperCollins*Publishers*

Thorsons
An Imprint of GraftonBooks
A Division of HarperCollins*Publishers*
77-85 Fulham Palace Road,
Hammersmith, London W6 8JB

Published by Thorsons 1987
3 4 5 6 7 8 9 10

A CIP catalogue record for this book
is available from the British Library

ISBN 0 7225 1423 9

Printed in Great Britain by
HarperCollinsManufacturing, Glasgow

Contents

NOTE TO READERS

Before following any self-help advice given in this book readers are earnestly urged to give careful consideration to the nature of their particular health problem, and to consult a competent physician if in any doubt. This book should not be regarded as a substitute for professional medical treatment, and whilst every care is taken to ensure the accuracy of the content, the author and the publishers cannot accept legal responsibility for any problem arising as a result of experimentation with the methods described.

CHAPTER 1

Introducing a Few People

Before I started writing this book I came across four people with a problem. It was a problem in which their doctors had not been able to help very much, although they had tried in all sorts of ways.

Peter

Peter was a successful architect. He worked hard. Too hard, his friends said. He also played hard and had become, in his late thirties, an addict to sail planing. Both he and his wife Beth had mastered the art of the elaborate little skimming board-like craft and at weekends, or whenever they had a chance, they would chase off in their car with their sail planes strapped to the roof to lake or sea. Elaborate wet-suits allowed them to keep up the sport for most months of the year.

Then two things happened almost simultaneously. Beth became pregnant and Peter started to get 'funny turns'. The first one of these happened in the middle of the night after a pleasant social evening at his sailing club headquarters. Peter woke up in a 'bit of a sweat' with a feeling that something terrible had happened. For the first time in his life he could actually hear his own heart beating. It was racing at around 120 beats a minute. Beth, who had been a nurse before they got married, anxiously took his pulse about four or five times before calling the doctor.

When the doctor came, he took an ECG tracing. It showed no sign of a heart attack. But the racing pulse persisted. Gently, he applied finger pressure to Peter's eyeballs. Then he did the same thing to the arteries in Peter's neck. Still the racing pulse and the feeling of foreboding persisted. He administered a sedative injection and asked Beth to phone him in an hour if the attack, which he diagnosed as paroxysmal tachycardia, persisted.

In the early hours of the morning, Peter's pulse was still racing and he was admitted into hospital for investigation. Once again, ECGs showed no abnormality except the racing pulse. Then suddenly it all stopped. Peter felt fine. Both he and his doctor thought that the whole episode of PT was due to a combination of too hectic a life, too much work, and too heavy indulgence in the shape of food and drink on the evening in question. Peter agreed to mend his ways.

Three weeks later, Peter developed another attack of PT at his office at about 5 o'clock in the afternoon. He had been lunching with a client but had felt fine and very relaxed. He remembered what his doctor had done and gently massaged the blood vessels in his neck. All of a sudden his racing pulse snapped back to normal. He did not tell Beth about it but reported the episode once again to his doctor who prescribed a drug called a 'beta-blocking agent'. This made Peter's hands and feet very cold and he could no longer enjoy his sail planing, so he stopped taking his tablets after a few weeks.

At 1 o'clock on a Saturday morning, Beth felt her labour pains starting. She was nearly three weeks overdue and they were both terribly excited. Baby was on the way! They had arranged for Peter to drive her to the maternity unit. But as he got in the car his PT started up once again. As he sat frozen with terror in the car Beth made a big decision. She ousted him from the driving seat and drove them both to the hospital, where she had a little girl and Peter was kept in for observation once again! This time, while he was sitting in bed fuming that he was missing out on the part he had planned to play in Beth's confinement, a young hospital resident seemed very interested in his case and delved quite deeply into the sort of food he ate. Once again, all Peter's tests were normal. But this time the young woman doctor in charge of the case suggested that she thought Peter's attacks might be due to magnesium deficiency. She prescribed a magnesium supplement. Peter takes it still and continues to do a lot of work and sail planing. He also spends a lot of time playing with his little daughter, Susan. To date he is free of any attacks of PT.

Rebecca

Rebecca is 35. She's married to John, a builder, and they have two boys aged 8 and 10. Rebecca described herself as a normal housewife. She rated her marriage to be happy and successful 'up until now' she said with a stifled groan and asked her doctor to give her 'something for her terrible pre-menstrual tension, otherwise she was sure that John would walk out one day'.

John really could not understand what was happening to his wife. 'For most of the time Rebecca is a delightful and loving wife,' he said, 'She's marvellous with the boys who are a bit of a handful at their ages. But then a few days before her period is due she changes — you would hardly know her!' John, who was an intelligent and caring man, described it as a sort of 'personality change' — 'One minute she's fine and the next — well, she can be a real bitch. She shouts at the children. The house, meals and everything go haywire — and as for me, my name's mud. I can't do anything right at all!'

Rebecca described herself as 'bloated and bitchy' at this time. Often she found herself dropping even her most treasured bits and pieces of china. Once she had nearly written off John's van while she was 'in one of her moods'. Then, as soon as her period started, she would be fine again. Her doctor had been really very helpful and sympathetic. He had prescribed 'water pills' for the bloating, and tranquillizers. Then he had suggested some high-dose vitamin B_6 tablets. Later, the contraceptive pill had been suggested. All this was to no avail and Rebecca asked to see a Harley Street specialist.

The specialist was marvellous and talked to Rebecca about her problem in depth for a long while. Rebecca found this very interesting but unhelpful as far as her symptoms were concerned. The specialist also suggested another sort of hormone — this time progesterone suppositories. They made her feel sick and so she stopped the treatment after a few weeks.

Her doctor had come across some research work that suggested that sometimes magnesium deficiency is associated with pre-menstrual tension and suggested that she take a magnesium supplement for a few weeks. As a result, Rebecca menstruated a couple of months later without any of her former problems. She now takes a magnesium supplement regularly.

Harry

Harry described himself as a 'twenty-year coronary cure'. He had had a fairly severe heart attack all those years ago and had followed his doctor's advice to the letter. He had given up smoking (which he had loved), taken up golf (which he hated). He had lost nearly 22 pounds (10kg) in weight and regularly took a drug that kept his blood pressure within normal levels. The only thing that worried him was his 'bumpy old clock, doctor'. He did not get any angina (that went with the smoking). But now and then, mostly at night, he would get an attack of severe palpitation. He had tried everything for it, peppermint, two extra pillows, two less pillows, eating at 5 o'clock instead of 7.30, a walk with the dog before going to bed — even not watching anything too exciting on evening television. He still got it and it disturbed his rest as all he could do was to sit up in bed, read and wait for it to go. His doctor diagnosed a type of paroxysmal tachycardia but made no suggestion other than taking a sleeping tablet, but this made him feel dopey for half the next day. In Harry's case, a friend had suggested he try some magnesium diet supplements. Harry now takes them instead of his 'sleepers', and sleeps marvellously without palpitation.

Adrienne

Adrienne had just suffered the worst tragedy a young mother could have. She had put her baby to bed at as usual one night. The baby had had the slightest sniff of a cold but had taken her feed well and had been absolutely fine as she tucked her up for the night. (Adrienne was still giving her a bottle at about 11.30, after which the baby would sleep the night through.) As soon as she picked her out of the cot next day, she knew it had happened. For a minute or two, however, she could not believe it. She tried to remember what she ought to do and then rushed over to her neighbour to phone for a doctor. An ambulance came in double-quick time but the ambulance men said it was no good. The baby was dead. A cot death, or sudden infant death syndrome, had struck another child down inexplicably and without warning. There was nothing she could do.

It took Adrienne a long while to have the courage to start another baby. When at last she did become pregnant again she was haunted by the fear that the same thing might happen. Everybody said it was a chance in a million and that lightning never strikes twice. Then she heard of a woman who had had two such tragedies. Panic-stricken, she approached her doctor — surely there *was* something she could do. Her doctor was a wise woman who had had four babies of her own. She told Adrienne that the whole business of cot death was a medical mystery and that she could not really suggest anything that was scientifically helpful.

She had, however, read that it had been discovered that quite often pregnant women eat less magnesium than they should. She also said that she had heard that perhaps this magnesium lack in the mother could result in a magnesium-deprived baby and that magnesium deficiency seems to be associated with the sudden infant death syndrome. She said that if *she* were pregnant again she would concentrate on high magnesium foods while she was pregnant or take a medically approved magnesium supplement.

Adrienne is doing just that and feels much better because she is taking a positive step in the way of health insurance for her unborn child.

As we look at the whole subject of magnesium in detail, some of the apparent *magic* of magnesium will seem to evaporate and be replaced by sound and straightforward scientific medical argument.

Whoever said that one swallow does not make a summer knew what he was talking about. As I write these words on a cold May day, I've just seen a handful of swallows wheeling and diving around my garden. They have not made the temperature any higher but their presence convinces me at least that there's hope of better weather on the way. This handful of cases are mere harbingers of many more that are making doctors look at magnesium in a new light of hope in difficult medical problems.

What has changed?

Our parents did not worry about minerals much — although the health record of the previous generation or two has not been brilliant. Our grandparents worried about minerals even less and our great grandparents not at all. Are we therefore getting ourselves all worked up into a great song and dance about minerals in general, and magnesium in particular, for no real reason? Personally, I don't think we are and the reason why I take this attitude is because of Change with an enormous capital C.

Like it or not, our world has changed out of all recognition from the world that we evolved to live in. But we pay lip service only to the new diseases of civilization. We are shy to speak up and make statements as to how they have come about by saying they are multifactorial and so we cannot do anything about them. Such an attitude is not only craven, it is also self-deceptive in the extreme.

Fundamental to good natural (and automatic) mineral nutrition is the very soil of our land. Modern intensive crop production repeatedly eases every possible bushel of grain out of every acre by the widespread use of more invasive ploughing, while artificial fertilizers leach out our mineral heritage. There is also evidence, all around our river estuaries, of lost topsoil being washed off and away down into the sea together with its minerals.

In the UK we do not seem to have woken up to this yet. In the USA, the US Department of Agriculture spelt their worries out to the public recently. They pointed out that five and a half billion tons of top soil were lost to the land in a graphic way by stating that for example, enough soil goes into the Mississippi River each year to build an island a mile long and 1¼ miles wide to a height of 2000 feet. It contains 800 odd railway wagons full of phosphorus 20,000 of potassium, 300,000 of calcium and, significantly for our interest, nearly 70 cart loads of magnesium too! And that's just one river in one country.

The other important factor is that today everybody — or almost everybody — is slimming. We have not learned the knack of matching our culture's food intake to reduced energy needs, due to reduced exercise and labour-saving devices that range from bicycles to washing machines, from lawn mowers to outboard motors, from electric drills to hedge clippers. The transition from walking to automobile travel has also made an enormous and permanent alteration to our lifestyle.

To counteract all these changes we and our children have to build calorie-saving methods into our lives to avoid getting obese. Often when folk decide to eat less, they eat less in the way of grain products and vegetables. But our bodies are designed around a physiological template that states that we need to eat quite considerable quantities of grain and vegetables to get all the minerals we need all the time. And so we see soil erosion and the modern crop-boosting economy depleting the mineral birthright of the crops in our fields and compounding the effects of altered dietary behaviour. As a result, modern man, woman and child are facing, probably for the first time in existence, *mineral malnutrition.*

This does not matter too much as far as some minerals are concerned, but in other cases it matters a great deal. For example, the trace elements such as selenium are extremely important for good health. And, of course, the health implications of a negative magnesium balance are extremely worrying too. Lack of magnesium contributes to a wide spectrum of health problems, as we shall see.

CHAPTER 2

The Minerals in Your Life

In many ways the substances we call minerals are the Cinderellas of nutrition. Everybody knows about the vitamins and the main building blocks of the nutritional castle (the carbohydrates, fats and proteins). But hidden away somewhere in the castle is Cinderella — being helpful, of course, in the nutritional kitchen. No one thinks much about her, but she provides an essential service — so essential that if her services are withdrawn the whole organization becomes disrupted and is liable to collapse.

In many ways, the word mineral is somehow old-fashioned. It was coined to describe a mixture of elements — and a 'dead' inorganic mixture to boot. This term has somehow or other clouded our appreciation of the importance of minerals ever since. We tend to look at them as though they were some sort of bone meal — a 'dead' amorphous substance if there ever was one. Of course, bone *is* largely composed of minerals. But minerals in our bodies are actively incorporated into living tissues and especially into enzymes and hormones — the very essence of life. If we ignore them in nutritional terms, then we do so at our peril, as we shall see.

To try to revitalize our knowledge of minerals and to understand their true worth and value — to take Cinderella out of the kitchen and get her to the Ball, you might say — the Food and Nutrition Board of the National Academy of Sciences in the United States has coined two brand new terms so that we can look at these vital substances in a new light and give them the stature that they deserve.

First of all, there are the *macro-minerals* (macro means large). These substances, so vital to our existence, are present in relatively large amounts in our body and so we need to include them in relatively large amounts in our diet. They are calcium, phosphorus and magnesium, together with sodium, potassium and chlorine — a strange collection of elements in many ways but all bound up together in the tissues of our bodies. Obviously, they do not occur in elemental form. For instance, phosphorus is so unstable that it has to be kept away carefully 'under wraps' or it immediately starts to burn into oxides and salts, and chlorine is a poisonous gas in its elemental form. Instead, these elemental macro-minerals exist in nature as compounds incorporated into complex organic chemicals that become part of life.

The other big group of minerals are the *micro-minerals* or, as they are more usually called today, the trace elements. There are rather more of these and they include iron, zinc, copper, manganese, iodine, chromium, selenium and molybdenum. In the United States another element, fluorine, is included as a micro-nutrient by most authorities. But although fluorine is useful in preventative medicine, in as much as it inhibits tooth decay, it is not an essential element to living matter. So, generally speaking, it is relegated to a limbo of lost souls in the underworld of nutritional science.

A question of size

It therefore seems to be a good idea to divide up the vital living world of minerals by size. The micro-minerals, or trace elements are vital to our bodies in 'parts per million' or 'parts per billion' quantities. Most people find it difficult to visualize such tiny amounts. One reasonable analogy is to imagine a wide-bodied jet aircraft filled up with the 14,000 gallons of fuel necessary to take it across to the other side of the world — and then to take a spoonful or so out! If this happened the aircraft could just fail to make its destination! This is what happens to us as far as trace elements are concerned. That little trace is vital to us, a part per million though it may be. If we have not got it we 'don't make it' nutritionally in the way that we should and therefore we tend to fall ill.

With our macro-minerals the relationship between necessary dose and health is not so critical. But the minerals are no less vital and size comes into things here too. An important 'how much?' question still has to be posed. It is possible to depict this in the form of a graph.

From this you can see that there are three important ranges for us to consider. Range 1 is bad news as far as nutrition is concerned because we are not giving our bodies as a whole enough of the mineral in question

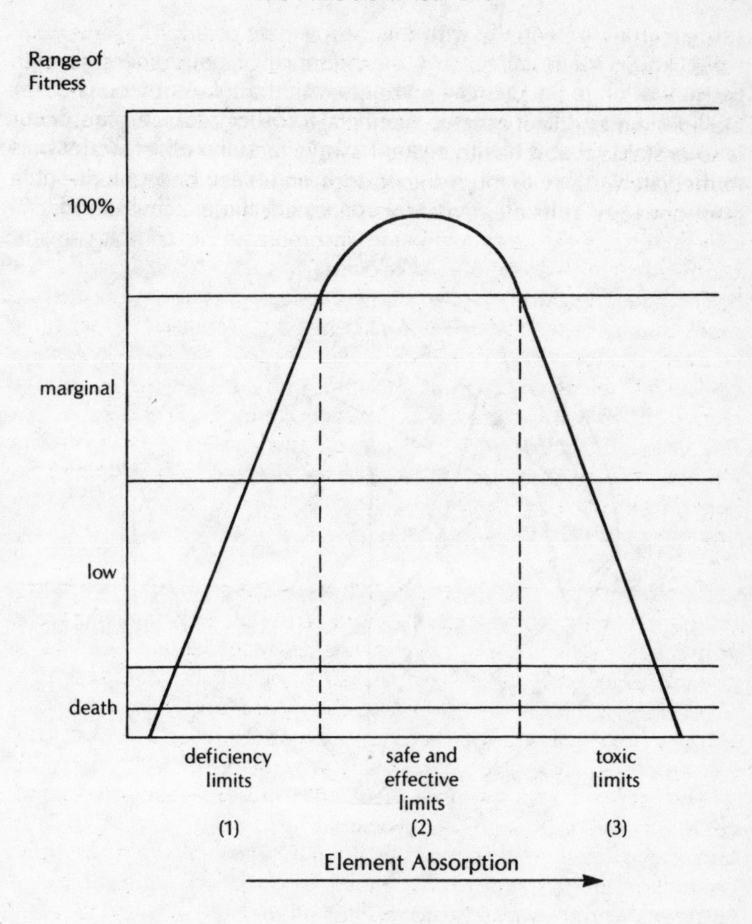

Figure 1

to keep us fit and well, and symptoms signal this to us sooner or later. Range 2 is where we would like to be. We are taking enough on board to keep us fit and well and you can see by the shape of the curve that a fairly wide variation of intake allows for a wide natural safety margin. Nature has planned it this way because there are all sorts of things that can go wrong potentially with this happy state of affairs.

We know, for instance, that the presence of one mineral in high concentration in the diet can compete with that of another one as far as the body's utilization is concerned. Another problem can occur because everybody's facility to absorb minerals (and other nutrients as well) is very variable. Some of us just lap them up. But I may have trouble absorbing, say, calcium while you absorb calcium easily but run into

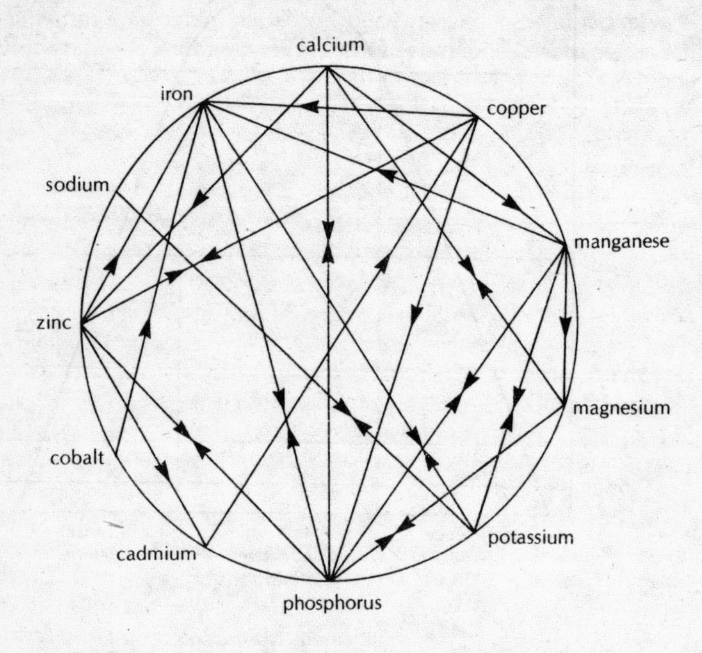

If you want to know if one mineral in your diet interferes with another, look at the arrows. If an arrow from one mineral points at another mineral, then a *deficiency* of that mineral may be caused by an excess of the mineral from whence the arrow originates.

(after Ashmead's Mineral Interference Wheel)

Figure 2
The Mineral Circle of Health

trouble over phosphorus. And so Nature designs us with this broad base of safe intake levels. In most minerals there is a third range. Range 3 is the toxic or poisonous range. It may come as a surprise that something as natural as a nutrient, a food, can be toxic, but it can be.

Having met some of the mineral family by name it is necessary to get to know them a little bit better so that we can get them into perspective, for they all have a function to play and in most cases there is a deal of interaction between them. So to understand something of the magic of magnesium we must know something about its brothers and sisters. But before we do this, I must introduce a technical term beloved by nutritionists — the Recommended Daily Allowance, or RDA. Broadly speaking, we need the RDA of every nutrient to keep us in Range 2 on our graph.

A question of quantity

It may come as a surprise to you but even in this day and age, when we can stroll on the moon or send back incredible close-up pictures of faraway stars and planets that were just bright spots in the heavens to a previous generation, we still do not know exactly how much of the very many nutrients we should take as our 'daily bread' with a view to optimum health. But we have made a brave guess at it!

The Food and Nutrition Board of the National Academy of Sciences in the USA has developed the most accepted guesstimate to date in its Recommended Dietary Allowances (our UK versions are only slightly different). But mathmeticians look sideways at these intelligent guesses for two reasons. First of all, they assume that nutrient requirements in normal healthy people are distributed according to typical 'Gaussian' distribution patterns (when everybody knows they are not!) Secondly, the assumption has been made that people deviate in their nutritional needs by only about 15 per cent of a given mean. This second assumption was arrived at largely as a result of experiments on the protein requirements of college students. It had practically nothing to do with other nutrient requirements at all and so is particularly suspect.

The Board admit the inborn errors of their guesstimates and agree that they involve a 20-30 per cent risk of total dietary deficiency! Before we look at minerals in general and magnesium in particular in this RDA context, it must be mentioned that there is a wealth of hard scientific evidence that RDAs can be very misleading. An experiment quoted by

Dr Jeffrey Bland in his book *Medical Applications of Clinical Nutrition* (Keats, 1983) demonstrates this particularly well as far as Vitamin C, a common enough nutrient, is concerned.

One way in which you can be sure that somebody is fully 'topped-up' with vitamin C is to find out how much vitamin C that person has to swallow before 50 per cent of what he has eaten is excreted in his urine. Nine normal adult females agreed to take part in such an experiment and the results were quite amazing for it was shown that they needed doses ranging from 0.6 to 2.2 milligrams of vitamin C (per kilogram of their body weight) to achieve this 'topped-up' 50 per cent excretion rate.

Experiments like this are difficult to design and carry out on large numbers of humans. But similar results for animals relative to vitamins and minerals underline this wide variation of nutritional requirements, and must make us look at RDAs, merely as helpful guesstimated guidelines. Convincing scientific evidence of our *real* nutritional needs for *optimum* health still eludes us.

Calcium, the mineral foundation

There is more calcium in the body than any other mineral, and 99 per cent of it is found in the bones and teeth, where it not only forms the mass of these organs but acts as a calcium store or reservoir. We know a great deal about how the body regulates its calcium blood content by means of a special hormone system which keeps about 10g of calcium 'liquid' in the blood and tissues, leaving over a hundred times as much 'on deposit' in the bones and teeth.

The RDA of calcium is 800mg and most nutritional sources suggest that we do not get enough calcium in our diet. Richest sources are milk, cheese and leafy green vegetables. Calcium is linked with the magic of magnesium, as we shall see presently with reference to the stability of the heart's rhythm.

Calcium deficiency erupts as symptoms in the form of muscular aches and pains, irregularity of the heart's action, osteoporosis (brittle bones) and periodontal disease (gum recession and tooth loosening.) Weight-reducing diets which are low in milk and cheese, the most common sources of calcium, can bring about deficiency problems.

Phosphorus

Next to calcium, phosphorus is the macro-mineral of which the body has the greatest need and it accounts for around 1 per cent of our total body weight. Bones have about half as much phosphorus in them as they do calcium and, like calcium, there is a constant hormone-based regulated turnover. As well as being a 'framework and strength' mineral, phosphorus is important to maintain energy release at a cellular level throughout the whole body. The RDA is 1000mg, but phosphorus is so readily available in meat, fish, poultry, eggs, legumes, milk and cereals that no one really goes short.

The electrolyte minerals

This 'electric three' — potassium, sodium and chlorine — are constant reminders of our historical 'fishy' heritage. In many ways, they produce a private piece of the ocean that we all carry about with us as a throwback to our aquatic beginning in evolution. Known as electrolytes, because they readily conduct a current of electricity, these three elements are, in many ways, life's blue-print in the animal world. Potassium and sodium, combined with chlorine (as chlorides), exhibit a strange tug-of-war relationship within our bodies and a complex and always changing relationship evolves around concentrations of the two elements. It is worth understanding about this because it is very relevant to our health and general well-being.

If you look at food from the point of view of an omnipotent designer, you can argue that we evolved as animals to eat food containing much more potassium than sodium. But our modern diets in fact contain more sodium than potassium, which is strange when you consider that our bodies actually contain twice as much potassium as they do sodium (9 oz/250g of potassium compared with 4 oz/100g of sodium). The relationship between sodium and potassium in the body is worked out very much around the conservation of sodium, which at the 'design stage' was much scarcer in our food than was potassium. Modern eating and food processing practice has changed this potassium domination, however, and we now manage to 'process' much more sodium into our bodies. Perhaps this is to our detriment.

Something like 98 per cent of the body's potassium resides *inside* our cells and this differs radically from the deposition of sodium in the body. This is largely extracellular. The sodium/potassium ratio *inside* the cells is also interesting. It is 10:1 while *outside* the body's cellular mass (for example, in tissue fluids) the ratio is 28:1 — a vast difference in gradient.

Our kidneys work hard to maintain this intricate balance. When we eat foods containing a high potassium content, although we absorb about 90 per cent of the ingested potassium, the blood level of the element remains constant — a remarkable feat of efficient excretion in action. Potassium and magnesium (the magic of which we start to examine in detail in the next chapter) are very much involved in the regulation of the heart's rhythm. Potassium deficiency was identified as a culprit in a disorderly action of the heart which occurred during the Apollo 15 astronauts' lunar exploration flights involving David Scott and James Irwin. By the time that the Apollo 16 astronaut exercise was mounted, potassium-enriched food was incorporated in the foods on this and all subsequent astronaut programmes and no untoward depletion symptoms developed.

From the practical point of view, people who are receiving medical treatment with diuretic drugs can run into trouble through potassium loss. These drugs are the 'water pills' that some patients have to take if they are to avoid 'water logging' of their tissues, possibly as a result of heart failure, when tissue fluids accumulate in the legs or abdomen, causing swelling or dyspepsia, or in the lungs causing coughing and breathlessness. (There are, however, more benign causes for all these symptoms.) Unfortunately, most diuretic drugs used to treat this sort of problem cause excessive potassium loss from the body and a disturbance of the sodium/potassium ratio *unless* adequate potassium supplements are taken.

Potassium deficiency

Although most potassium deficiency is caused by medical treatment, there is increasing concern about the effects of low potassium intake in modern diets. As already mentioned, we were 'designed' for much more potassium than we get these days and deficiency symptoms include abnormal tiredness, irritability, headaches, swelling of the feet and ankles, palpitations and bone and joint pains.

The much favoured processed foods many people eat these days are bad news to potassium intake. They are also rich in sodium which tips the sodium/potassium ratio off balance and, although potassium is plentiful in many fresh fruits and vegetables, it is almost entirely lost by canning and freezing processes.

It is impossible to discuss the electrolyte minerals without entering into the salt/blood pressure debate or, as many would say, controversy. High salt intake is sometimes associated with high blood pressure but it would seem that not everyone is sensitive to salt in this way. A certain number of folk with high blood pressure do manage to normalize their blood pressure by adjusting their potassium/sodium intake to the sort of ratio that we were all designed to cope with. (It is important to realize that the normalization of high blood pressure is always a matter for the doctor rather than the patient to deal with.)

Potassium and energy is another fascinating aspect of mineral health control. Without adequate cellular potassium the body has difficulty in converting blood sugar into glycogen, which for all intents and purposes can best be looked upon as an instant energy 'storage pack' that can be converted into energy very rapidly. In the absence of adequate potassium, our blood sugar tends to rise and this in turn stimulates a need for extra insulin to be secreted, and of course the diabetic process is essentially one of increased insulin demand.

The magic of magnesium is also closely tied up with potassium on the energy front, providing yet another link between magnesium and the electrolytes.

How much potassium is enough?

The 1980 RDA suggestions recommend maintaining the current average intake in our diets (2000-6000mg daily) while reducing the sodium intake by half. Apart from those who, for reasons of poor absorption or diet manipulation, tend to need supplementary potassium, the important way to play it safe and healthy as far as electrolyte intake is concerned is to 'hunt and destroy' unnecessary and hidden sodium in our food. A sensible balance should be struck here. The most obvious thing is to avoid added table salt and to eat with discretion when it comes to canned or frozen vegetables, cured, smoked and canned meats, fish and sausages, peanut butter, potato chips, salted nuts and crackers, canned or packaged soups, processed or cheddar cheese, sauerkraut, and large quantities of olives or salad cream.

The trace elements vital to fitness

Having lightly 'pencilled in' the general design of the main fabric of the 'magic of magnesium' that forms the main substance of this book, it is necessary to mention in passing the micro-minerals. (They are so classified because of the relatively small amounts in our food.) It is necessary to state, as loudly and as clearly as possible, that these 'traces of excellence' — the trace elements — are enormously important.

The famous five

1. Iron — the medical mineral

Iron is the most vital component of our red blood cells — all 20,000 billion of them. It may come as a surprise that we classify iron as a trace element. And yet it is one. Your whole body contains only 3.5-4.5g of it — the amount of iron in a builder's nail. The reason we need so little iron is due to a miracle of conservation, a feat of recycling that a man-made chemical recycling plant can in no way match.

Once it has been absorbed, the body is very 'mean' with iron. Fit men lose iron mostly due to cell loss from the skin and the bowel. Women lose much more iron during their fertile lives through menstruation and having babies. Very little iron is ever excreted and the body regulates its internal mineral iron by an incredibly sensitive selective absorption process.

Normally we absorb about 6-10 per cent of the iron in our food, but an anaemic person can absorb up to 60 per cent. Nobody understands how the absorption organ *knows* how much iron to absorb to keep the balance right. But it does.

Iron deficiency is the commonest mineral deficiency disease and US Health and Nutrition studies demonstrated that 95 per cent of children between the ages of 1 and 5, and most women during the fertile years, have an intake of iron below the RDA. In adult men, the RDA is 10mg per day. Women need 18mg per day during their fertile years.

The management of iron deficiency (anaemia) is basically a medical problem but it is possible to improve iron nutrition by means of suitable diet manipulation.

2. Zinc — the mineral that wanted a measuring stick

Doctors and nutritionists have only recently started to be interested in zinc as a mineral nutrient and there are two very good reasons for this state of affairs. First of all, there is so much zinc around in the soil that everyone assumed that foodstuffs would be pretty well packed with the stuff. Secondly, until a comparatively new measuring tool, called the atomic absorption spectrophotameter, arrived on the scientific scene quite recently, analytical procedures for measuring zinc in the body were both difficult and inaccurate.

Experiments in 1974 first of all suggested that zinc might be a member of that exclusive health club of essential mineral nutrients and in 1974 the National Academy of Sciences added zinc to their RDA list. Zinc might well be considered to be the *enzyme* vitamin for it is especially important in three major enzyme systems, one involved in bone metabolism, one involved in pancreas function and another very much bound up in connective tissue maintenance. But zinc is also involved in sperm function and the efficient breakdown of alcohol in the body.

Our bodies only contain about 2-2½g of zinc and so in fact the body's reserves of zinc are always small. Symptoms of zinc deficiency include loss of appetite, skin disease, poor healing and growth and, perhaps strangely, loss of the sense of taste. Zinc is well absorbed from food by most people and is abundant in most diets.

Zinc deficiency only seems to occur in special circumstances. First of all, there are folk whose absorption organ seems to be deficient as far as zinc is concerned. Then there are certain foods that interfere with zinc absorption — particularly fibre. Thus high-fibre diets can lead to low zinc levels in some people. Low calorie diets and highly refined foods, like white flour, are zinc-depleted. In some areas of the globe, zinc-depleted soil and crops have been reported.

A well-balanced diet should supply the normal fit healthy person with the 12-15mg which RDA studies suggest is adequate. Doses of 50 to 75mg per day are often given to those suspected of zinc deficiency.

3. Copper — the mystery payout

In many ways, copper remains a mystery trace of excellence. Like zinc, it is present as a building brick in many enzymes. Our bodies only contain about 100mg of copper and it is divided three ways. A third resides in the liver and brain, a third is in the muscles, and the rest is fairly evenly scattered through the body. Copper deficiency is suspected as being involved in a whole host of human ills that range from heart disease to schizophrenia and from pill (oral contraceptive) induced depression to arthritis.

An interesting aside on copper-lack as related to rheumatism came to light as a result of the copper wristlets and bangles so popular during the 1970s, when it seemed that almost everyone you met wore such a rheumatism band. Doctors in general and rheumatologists in particular were almost certain to smile indulgently on spotting such a trinket on one of their patients. That was until it was demonstrated that a measurable amount of copper 'leaves' these rheumatism bands day by day and is absorbed by the skin.

An Australian rheumatologist, Dr Ray Walton of the University of New South Wales, Australia had, like most doctors, long decided that the undoubted improvement in many of his patients' rheumatism when they wore copper bracelets was due to placebo response (placebo = I please). But was it? He had some bogus copper-coloured bracelets made up of aluminium alloy and issued these to sufferers as traditional copper bracelets. Again, there was some improvement (showing the placebo response to be alive and well and functioning, as it always does). But far fewer of the alloy-bracelet wearers responded favourably than did the wearers of the real copper bracelets. This suggests that a mineral effect in excess of the placebo response was making itself felt.

Copper is pretty ubiquitous in our food and we need about 2-3mg per day to keep us well topped up. There is evidence from the United States that soil copper and, therefore crop copper, is falling. Oysters, nuts and liver are copper-rich foods. Deficiency effects are pretty rare, it is thought, except in poor absorbers who perhaps develop rheumatism.

4. Manganese — the trace mineral that resists absorption

In all probability we shall learn a lot more about manganese as time goes by. Manganese is far harder than the other trace elements for the body to absorb, and this makes it difficult to keep the manganese balance right. In fact, we need very little manganese — perhaps as little as 10mg daily, most of which we store in bone. Within our bodies there is a complicated manganese cycle in operation.

Manganese lack has been related to schizophrenia, epilepsy, various metabolic diseases and possibly cancer and heart disease. There are examples of manganese being used as a medicine but sluggish absorption poses clinical problems. The RDA of manganese is 2.5-5mg per day — a pretty worrying level when potential absorption problems are considered. Many nutritionists would increase this RDA boldy. The main dietary sources are leaf vegetables.

5. Iodine — the trace element that started it all

Goitre — a swelling in the neck of a thyroid gland that tries to work harder to manufacture its essential hormone in the face of dietary deficiency in iodine, was endemic at the birth of the century in parts of Britain, Europe and the USA. Today it is a rarity due to another rarity — medical interest in trace elements! It was found that by giving people a little extra iodine in their table salt, goitre could be abolished.

The body contains 20-30mg of iodine, most of which is in the thyroid gland though the blood and tissues contain minute amounts of the element. Current interest in iodine has centred around radioactive fall-out which contains radio-active iodine. If the thyroid gland is deficient in iodine, the radio-active form 'homes-in' on the thyroid where it can produce thyroid cancer. Thus potassium iodide was issued to Chernobyl victims.

To prevent goitre, we need to take between 50-70 micrograms of iodine per day. Due to our relatively high salt intake, most of us get more than enough. Those on salt-free diets find that fish, liver, eggs, wholemeal bread and dairy produce provide adequate intake.

Ultra-trace minerals

There are three of these. Selenium is extremely important. The other two, chromium (yes, the metal they plate on to bicycle handlebars) and molybdenum, although clearly important to our health and well-being, are somewhat beyond the scope of this book but are mostly involved in the maintenance of blood sugar levels and enzyme stability.

CHAPTER 3

The Mineral with the Magic

What is it that makes magnesium so very special in the family of vital minerals that serve our bodies? To start with, there is not an enormous amount of magnesium in us — about an ounce, to be precise. This makes it the fifth of the five minerals we have briefly examined, after calcium, phosphorus, potassium and sodium. When most people think of magnesium, they usually remember magnesium ribbon, that aluminium-coloured element that burns with a bright ultra-violet flame at the touch of a match. It is also the silvery wire you see in a photographic flash bulb. Of course it enters our body, not in this metallic form, but in the same way as do most other minerals in our food. That is, as a compound (sometimes called a salt) combined with other elements.

About 70 per cent of the ounce or so of our tissue magnesium resides in our bones and teeth, but once present in these organs it appears to be well 'locked up' or 'locked in'. In other words, there is little or no interchange between the magnesium in our bones and teeth and the tissues of the rest of our body. This means we rely very much on a *regular day-to-day* intake of the element to make sure that our needs are well and truly met.

If you read medical textbooks you will find little space given over to talk of magnesium deficiency. There are two main reasons for this. Firstly, magnesium deficiency disease is very rarely looked for, and this is largely due to general medical apathy. Secondly, there is no really reliable and *easy* way to measure magnesium deficiency by scientific tests. I do not mean that there are no tests. There are several. But one snag is that, although the body keeps its *blood level* of magnesium remarkably constant, this does not mean that the actual *tissues* of the body contain sufficient magnesium to function optimally.

Scientists have tried to measure tissue or *cellular* tissue magnesium in many ways. First of all they tried to do so by estimating the amount of magnesium in the actual red blood corpuscles. This has proved to be rather unsatisfactory. More recently, there has been a move to measure the magnesium in the white cells (leucocytes) but this is technically difficult too. Another way is to measure the magnesium in actual muscle by doing what is called a *punch biopsy* (in which a little bit of muscle is removed from the body for analysis — an uncomfortable procedure).

Probably the only really efficient way to find out if someone has enough tissue or cellular magnesium to ensure good health and to exclude the possibility of a magnesium deficiency causing symptoms is to carry out what is known as a magnesium load test. This involves collecting the patient's urine over twenty-four hours on at least two occasions and determining how much magnesium they excrete after an intravenous 'loading' injection of magnesium has been given. This is a laborious, expensive and unpleasant sort of test, you will agree. No wonder doctors and laboratory workers are not keen on obtaining scientific back-up for the magnesium 'status' of their patients. Despite this reluctance, however, some scientific information has been obtained on magnesium deficiency by laborious research using this sort of test.

Deficiency indication by symptoms

Understandably, however, most of the information on magnesium deficiency has accumulated in a rather different way than by laboratory tests. In other words, it has come about by looking at animals — and that includes people — who take a magnesium-deficient diet and noting the symptoms they develop as a result of their malnutrition. When you consider that magnesium deficiency symptoms may include loss of appetite, palpitations, irritability, weakness, insomnia, muscle tremor, numbness and tingling sensations, confusion, personality change and skin problems, you can appreciate that the 'magic' that occurs when people get relief from a magnesium supplement does seem really pretty dramatic.

But, as we will see in later chapters, it is not only troublesome symptoms that tend to evaporate when the hidden deficiency of magnesium is made good. The whole progress of a particularly nasty gaggle of diseases can be influenced too and the first of these will be examined in the next chapter.

How much is enough?

The current RDA as stated by the National Research Council is 350mg for men and 300mg for women. The United States estimate of the FDA (Federal Drugs Administration) is 400mg. Most magnesium experts like to link RDAs of magnesium to a dose that is related to the individual's weight. This works out at a level of 6mg per kilogram per day.

There is a fair amount of evidence that large numbers of people do not get enough magnesium in their everyday life.

In the United States there have been magnesium watchers around for a very long time — since the 1930s, in fact — and they have recorded, according to Dr Mildred Seelig of the Goldwater Memorial Hospital, New York University Medical Centre, a gradual fall of the magnesium content of the national diet. Now, one interesting thing about magnesium is that certain nutrients and foods *increase* the body's *requirements* of magnesium — notably vitamin D and also phosphorus.

There was a sharp rise in all our intakes of vitamin D in the 1940s and 1950s, with the fortification of various foods, especially margarine. Shortly after this, there was an 'explosion' in the soft drinks industry on both sides of the Atlantic. The consumption of soft drinks provided us with a major source of phosphorus in our diets, and the rate of consumption has ben rising rapidly for the last quarter of a century.

And so we have a very interesting and maybe worrying situation in which, not only are we eating *less* magnesium, but we are also eating much *more* in the way of foods and drinks that demand *more* magnesium to keep us well topped up with the vital element.

Incidentally, if you and I are putting ourselves in a difficult position as far as magnesium is concerned, then our children are even more likely to be in jeopardy. Thirty years ago, the British Paediatric Association pointed out how much more vitamin D we were giving our children each day — especially when we unconsciously add to it with a diet supplement. For instance, a toddler taking 1½ pints of dried milk per day together with an ounce of cereal and a teaspoonful of cod liver oil could tot up a daily total of over 4000 international units of vitamin D daily — far more than was necessary to protect him from rickets and enough to point the child in the direction of the magnesium deficiency syndromes which we will examine presently.

The magnesium intake of young people

Another group's dietary habits have quite recently been analysed and reported in the *Journal of the American Dietary Association*. It involved looking at what students ate and drank each day in fifty colleges in the USA. It was a momentous task and the findings were quite worrying from the point of view of magnesium intake. This averaged a mean of only 250mg per day (RDA = 400mg per day). What was more, the phosphorus magnesium ratio was 7:1 — a high ratio, and you will remember that the more phosphorus you take the more magnesium you need.

Another change of lifestyle that most western countries enjoy these days is an increasing intake of alcohol by both young and old. It has been amply demonstrated that even moderate social drinking is what the nutritionists call *magnesuretic*. In other words, magnesium is pumped out through the kidneys. No wonder so many middle-aged folk find that they can no longer take a whisky or a gin as a pleasurable addition to their evening meal if they want to have a night free of palpitations. That extra drink can force just that little bit of magnesium out of their tissue fluids and into their urine to put them out of comfortable magnesium balance and their heart starts to register its complaint by over-acting for an hour or so.

So it rather looks as though whenever and wherever you look for evidence of magnesium deficiency in the diet you tend to find it and women seem to be particularly at risk. A recent survey of pregnant women from different economic backgrounds showed that their magnesium intake varied from 103 to 333mg per day (averaging out at around 204mg). Another group of doctors writing in the *American Journal of Clinical Nutrition* carried out meticulous seven-day metabolic balance studies of ten healthy white women in Tennessee and found that their mean daily magnesium intakes were only 60 per cent of the recommended pregnancy RDA of 450mg/day. Some of the possible risks that women run from low magnesium intake during pregnancy are dealt with later and Chapter 5 deals with the special magnesium problems of women.

Another cause for concern as far as magnesium deficiency is concerned is that nowadays everybody is tending to eat much more fibre. In many ways this is an excellent health trend, but there is one rather awkward fly in this otherwise encouraging therapeutic ointment — high fibre, brown bread, brown rice, oatmeal, muesli, or phytate-enriched white bread all tend to reduce magnesium absorption.

All in all, the consensus of nutritional opinion on both sides of the Atlantic is that the magnesium supplied by the average diet is at best only marginally adequate and that all sorts of things are liable to tip the balance the wrong way and make for a shortfall. It could be the sort of bread you eat, whether or not you do a heavy job or play an exhausting game that makes you sweat a lot. It could be a question of sex (men tend to get depleted of magnesium more easily than women). Genetic factors, which in turn may decide how efficient or otherwise your absorption organ is at leaching the magnesium out of your food, come into it too — and so does where you live and the hardness of your water supply.

It has been calculated that about 12 per cent of our daily intake of magnesium can be derived from water and that if you live in a hard-water area you can get 18 per cent of your magnesium via your water supply.

Rich in magnesium

A list of magnesium-rich foods makes a curious mixture. Cocoa or chocolate contain a lot of magnesium, plain chocolate more than milk. Nuts are good sources, too, especially cashews, almonds and brazil nuts. Seafood generally is an excellent source, especially winkles, whelks and shrimps. Some vegetables are pretty rich, particularly beans, peas and spinach beet. All grain is a good source, with barley and wheat topping the poll. A diet is of course a mixed thing and the sort of general foodstuffs that we tend to eat in *quantity* are very poor providers of magnesium. Meat and fish contain only relatively small quantities of magnesium and the same is true of fruits, salads and dairy products.

Perhaps by now you can see who is most likely to be self-selecting a really low magnesium diet. Of course it is the person who is overweight and therefore decides to slim by opting for a high-protein, low-carbohydrate diet.

Certainly the person on a steak and salad routine will lose weight in the fullness of time, but his magnesium intake plummets instantly too. This is particularly unfortunate if the slimmer has been told by his doctor to lose weight because of a heart condition for, as we will see in the next chapter, healthy hearts need the magic of magnesium.

High Magnesium Content Foods
(mg per 100g edible portion)

Food	*mg*
Wheat bran	550
Cocoa powder	450
Winkles	400
Brazil nuts	400
Cashew nuts (roasted)	265
Soyabeans	250
Soya flour	250
Baker's yeast (dried)	220
Almonds	200
Peanuts	180
Whole wheat	175
Beans (dried)	170
Pistachio nuts	160
Brown rice	160
Hazel nuts	150
Whelks	150
Walnuts	130
Oats	130
Maize	120
Peas (dried)	120
Barley	105
Milk chocolate	105
Plain chocolate	100
Shrimps	70

CHAPTER 4

Magnesium and Your Heart

Sometimes it is possible to understand health subjects quite easily because the cause and effect principle involved is straightforward and simple. For instance, if you stop smoking and cease coating the cells inside your lungs with tar every day of your life, you are less likely to develop cancer of the lung. Similarly, if you change from heavy drinking to a moderate intake of alcohol, it is less likely that your liver will complain and develop cirrhosis.

But with heart disease the subject is not so simple — although it is not really all that complicated either. Basically, the heart is a pump, and all pumps have two things in common — valves and a power source. Valves are essential to any pump in order to direct the flow of what is being pumped. Some years ago valvular disease of the heart was a common enough illness. Nowadays, due mostly to the invention of antibiotics, valvular disease of the heart is becoming much less common and it is cardiac power failure that produces most heart problems.

Power failure in the heart really comes about as a result of muscle failure, for the heart's muscle is the sole source of its power and force. Muscles need nutrients if they are to work efficiently and these nutrients come in three main forms. First of all, there is oxygen, the prime nutrient of all our cells, including muscle cells. Secondly, there is a need for fuel (as glucose) which you might say is the heart's petrol. The third class of nutrients essential to a hard-working and efficient heart are our old friends, the essential minerals.

Nutrient failure equals heart failure

This relatively simple statement is true enough but it needs qualification. A pump can fail in all sorts of ways. If you ask it to do too much it will fail. For example, most people are familiar with the little miniature pumps that circulate water around a central heating system. They are designed to move water through a system that has various elements in it (radiators and tanks) that are at differing levels of, say, 20 feet. In other words, the boiler may be in the kitchen and the highest point in the system is up in the bedroom. If you put a small domestic pump of this kind in, shall we say, a ten-storey hotel or office building central heating system, it will do its best to circulate the water but the task will prove too much for it and it will fail because it cannot pump at a high enough pressure.

A similar state of affairs happens with the heart when the disease we call blood pressure develops. This is rather like having a central heating system installed in one flat or apartment and then connecting it to the rooms on the next storey above, and then, perhaps, to the one above that. Quite soon, if you did anything so foolish, somebody would complain that the radiators were not warming up due to poor water circulation. The heart differs from a central heating pump in that it will try to do any task you ask of it. Add some extra 'radiators' and it will gradually try to 'grow' into the job. But keep on taxing it and it will surely fail and the patient develops, in the case of high blood pressure, the disease called hypertensive heart failure.

A more common form of heart failure might be described as a sort of primary power failure. If you cut off the supply of electricity to a central heating circulation pump, it stops and the system fails. The same applies to the heart. Although the heart exists to pump blood it also needs its own supply of blood ducted to it through the coronary arteries (so called because they wrap themselves around the heart like a corona or crown). Interference with a regular supply of nutrients through these coronary arteries either causes heart disease — for instance, the cramp-like chest pain called angina — or proves fatal if the interference is great enough to stop the pump running completely.

Interfering with your pump

Angina is usually caused by the coronary arteries becoming diseased and the name of that disease is atherosclerosis (commonly called hardening of the arteries). A characteristic of angina is that its cramp-like pain comes on when you exert yourself. The reason for this is simple. When you are not demanding much of the heart's pump action — when you are, say, walking down the road or sitting watching television, the heart can get enough nutrient (blood) to tick over nicely enough, even through a constricted artery. Give it a bigger task and it suddenly demands much more blood to bring it the oxygen and foodstuff it requires to function 'under load'. If the coronary arteries are too small in 'bore' to supply this extra blood, angina develops.

The big sudden interference with cardiac function that is sometimes fatal is the coronary thrombosis. In this case, a clot (or thrombus) forms in a coronary artery, blocks it and causes an acute lack of nutrient, especially oxygen. Because of the way the heart muscle is supplied with arteries, a whole 'wedge' of the heart's muscle suffers (dies) when its coronary artery is blocked, for there is little alternative (co-lateral) blood supply in the heart.

Ten years or so ago, this general description of what can interfere with your heart as a pump could finish here. But more recently new knowledge about the parts that unsuspected nutrients (minerals) play in the heart's health and function have become manifest. In many ways this has been something of a potential therapeutic breakthrough, although — perhaps because doctors are very conservative people and get excited about new medical things rather slowly — not so very much has been heard on the matter.

The slow boat *from* China

The old song about the lad who loved his lass so much that he wanted to take her on a slow boat to China can be seen making a sort of return trip here! I would like to take you on a quick trip *from* China back to the West to demonstrate what seems to me, and to many other medical people, some very important news about cardiac health.

The dream of Klaus Schwartz was that as a result of his scientific endeavours he would live to see the eradication of just one disease. His desire was fulfilled because Dr Schwartz was instrumental in first telling the world about the microtrace element selenium, and in stimulating medical research on the part it played in human nutrition. Shortly after the discovery of selenium, a special breed of medical personnel, now called medical geographers, started to appear and began interesting themselves in the variation in soil content of selenium throughout the world. The amount of selenium in soil is closely connected to the amount of selenium found in the crops that are raised on that soil and subsequently in the animals (including human animals) that live on those crops.

But, to get back to China, for as long as medical records have been kept an unusual disease has been prevalent there, which has been responsible for the deaths of large numbers of children. To start with the symptoms of these sadly-fated children would be comparatively trivial but in the earliest stages of their illness it could be seen that they were not 100 per cent fit. They would gain weight slowly and be particularly prone to childhood illnesses. They would sweat a lot and would feed poorly. At meal times such children became irritable as they did their best to get food down before lack of energy made them too exhausted to eat. When doctors examined these children, they noticed a racing pulse, and a pot belly, due to an enlarged liver. Slightly older children suffering from this illness — now called Keshan disease — would get breathless easily and would suffer a hacking cough. As they coughed their neck veins would become very prominent. Special tests soon determined the basic cause for all their problems — they were suffering from a form of infantile heart failure.

Autopsy of those that died revealed that the muscle of their hearts, the tiny powerhouse of the pump, was flabby and contained widespread red patches. As luck would have it, one of the scientists involved in Keshan disease had veterinary knowledge and was struck with the similarity of the heart in this disease with an illness that killed poorly nourished pigs. The veterinary surgeon who had first researched this pig disease had thought that these red patches in the heart muscle of the sick pigs looked a bit like mulberry fruits and so he christened the disease mulberry disease. In time, it was found that mulberry disease only affects pigs raised on a diet that is deficient in the micro-mineral selenium.

Medical geographers knew by now that the Mianning county area in China, where Keshan disease was rife, was an area of low soil selenium and eventually in 1974, 4510 children born in this area were given selenium supplements to take. A further 3985 children were given a selenium-free dummy tablet to take as 'controls'. During the next year, 1.35 per cent of these controls developed classical Keshan disease while only 0.22 per cent of the children given selenium became ill. The next year it was thought to be unethical to deny controls a selenium supplement and soon the incidence of Keshan disease fell from 1.35 per cent to 0.03 per cent. Selenium supplementation is now universal in the Mianning area and Keshan disease is a rarity.

This story from China is quite fascinating, linking as it does a simple nutrient with a form of heart failure. The study raises all sorts of questions that are difficult to answer. For instance, why did a little over one child in 100 develop Keshan disease when they were all on a low selenium diet? Clearly there was something else about these children (perhaps they were poor selenium absorbers). The other thing is that, although there are other areas in the world with selenium-depleted soil, Keshan disease only occurs in China. But other research in the world of the medical geographers has linked selenium depletion and heart disease.

A trace element and heart attack

Anaheim in California is the headquarters of the American Society for Experimental Biology. Research workers there have been probing deeply into medical geography and have shown that Americans living in the selenium-deficient states of Connecticut, Illinois, Ohio, Oregon, Massachusetts, Rhode Island, New York, Pennsylvania, Indiana, Delaware and the District of Columbia have a heart disease death rate that is above the national average.

In fact, it is relatively easy to make a list of places throughout the world that link high or low coronary heart disease incidence with the amount of selenium that occurs in the diet. But the relationship between what we eat and how we get ill as far as the heart is concerned is not all that simple. Some of the earliest research on this subject was carried out many years ago by a well known British nutritionist, Dr Hugh Sinclair. He noticed a very strange thing about Eskimos living in Greenland. They very seldom, if ever, died of coronary thrombosis, but if they had to have one of their teeth out they bled profusely!

Dr Sinclair decided that Eskimos must eat something which interferes with the blood's tendency to clot. He persuaded his fellow researchers to eat and live like Eskimos for several months. He joined them in the experiment and they all ate masses of fish and seal meat. In the course of time their blood 'went Eskimo' too. It took very much longer to clot than it had done before they began their Eskimo diet.

This was really the start of a whole dietary revolution that has led nutritionists and doctors to look at the fat in our diet very carefully and closely in relation to the incidence of coronary heart disease.

In recent years there has been something of a campaign to persuade us all to eat much less fat to keep our hearts healthy, but this in itself has not been totally helpful in preventing coronary disease. Even though Dr Sinclair and his colleagues ate much *more* fat during their experiment, they ate themselves into a reduced coronary risk situation like the Eskimos. This was because the fat they were eating was a *special marine animal fat* containing natural anti-blood-clotting factors that helped to keep the Eskimos coronary-free yet liable to dental haemorrhage.

This nutritional message, concerning the type of fat that needs to be avoided, has been slow to sink in, but it is now being realized that just switching from butter to margarine is not going to be the answer in protecting coronary arteries.

A multifactorial puzzle

Unfortunately, death from coronary disease continues to represent a major cause for concern. Ischaemic heart disease (narrowing of the arteries) is the major cause of death in all affluent communities. There is an impression that death from coronary disease is a middle-aged hazard to a large extent because it is in middle age that so many die of the disease. But when circumstances allow post mortem reports to be carried out on large numbers of *young* men, a surprising number are already found to have coronary disease. For instance, of 300 young American soldiers killed in the Korean war, 70 per cent of the men between the ages of 20 and 44 had moderate to advanced arteriosclerosis.

Understandably, there has been a research explosion in trying to find the cause, and all that doctors have been able to say for sure is that it is a multifactorial disease, and that the factors include such things as smoking, lack of exercise, stress, diabetes, high blood pressure and perhaps improper nutrition, especially with reference to excesses of simple sugar and animal fat in the diet.

What is becoming obvious is that coronary artery disease is related to many factors other than those listed. For instance, as we have seen, it would seem to be related to the amount of selenium in the diet. But it is also related to the substances in the blood supplying the pump. The protective factors the Eskimos 'eat into' their cardiovascular system is a case in point. There is also another factor that looks relevant and has been largely forgotten. One thing that has puzzled heart specialists for some time is exactly why, for instance, Jack Jones dies in a few minutes with his heart attack while Harry Atkins, who may have a very similar heart attack, survives. An enormous amount of research has been carried out on this subject and perhaps the most promising lead to be followed recently involves magnesium.

—Coronary thrombosis with no thrombus!—

The American cardiologist, Dr W. Raab has made a name for himself as a tireless researcher into the whole field of cardiology and particularly with reference to the clinical features of what he calls 'so-called coronary disease'. He feels that the 'official' approach to much cardiac disease is based on outdated concepts. Dr Raab caused quite a stir at a symposium on myocardiology a few years ago by claiming that in about half the deaths clinically attributed to myocardial infarction, coronary occlusion, coronary thrombosis and coronary artery disease (all synonyms for coronary thrombosis), *no* thrombi (clot) or vascular occlusion (blocked artery) could be found at autopsy at all! And he put forward the theory that it was changes in the heart muscle chemistry that were really 'stopping the pump' in such cases.

Before we can understand Dr Raab's theory more clearly, it is necessary to examine something of the mechanisms of the normal heartbeat. For the heart to function efficiently it has to pump regularly and rhythmically. Unless it does that, the chambers of the heart cannot fill properly. This filling occurs when the heart muscle is 'resting' between beats. This regular rhythmic action is brought about by means of a complex conduction system within the heart's muscle — the 'wiring' of the pump, you might say. But there is another factor of great importance too. This involves what scientists call *ionic* considerations. It has always amazed me how science, and especially medical science, seems to try to make quite easy concepts seem difficult. The word *ion* comes from the Greek and literally means 'going'. In science, the term was coined to describe an electrically-charged atom. When chemical substances are described as being 'ionized' it means that an electrolytic dissociation has occurred and the atoms are capable of 'going somewhere'. This 'get up and go' phenomenon with reference to many of the minerals we read about in Chapter 2 is essential to another aspect of normal cardiac function for its seems to condition, or programme, the heart muscle itself to 'behave properly' and contract rhythmically and efficiently.

Dr Raab, like many of his colleagues, apparently could not miss out on an opportunity to sound a bit mysterious and learned. So he too coined a phrase to explain why so many people drop dead with coronary 'thrombosis' without an actual thrombus being present! He called it cardiac *dysionism*. (When doctors put *dys* in front of a word, it is shorthand for *difficulty*. So dyspnoea means difficulty with breathing, dyspepsia difficulty with digestion. Dysionism means difficulty with ions). Clearly, the difficulty seems to be particularly with the magnesium ions that are so vital for proper muscle functioning.

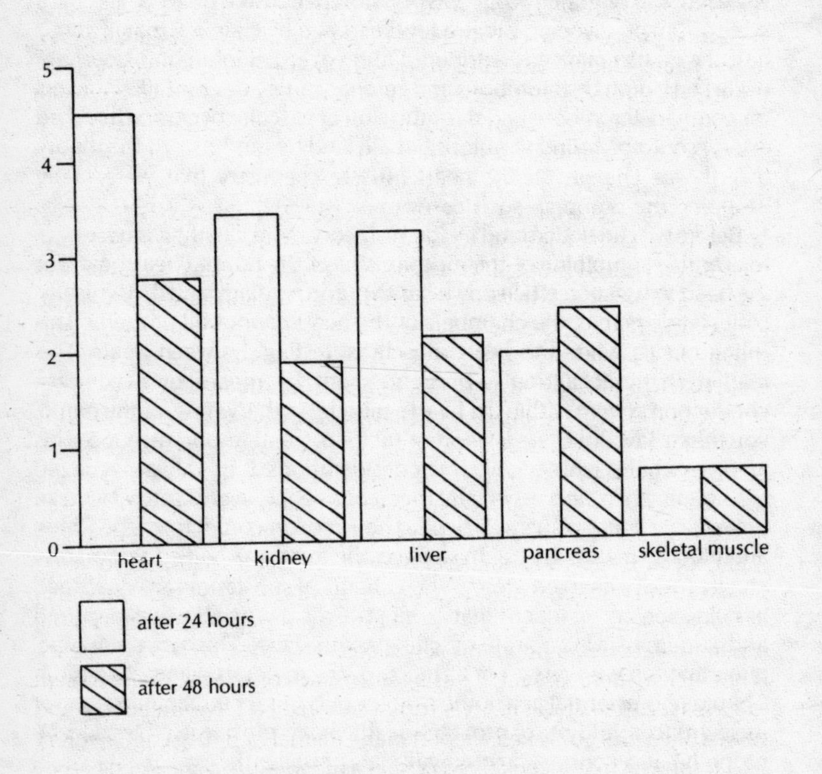

Figure 3
Where Magnesium Goes To
magnesium 'homes in' to the heart

The cardiac magic of magnesium

According to Dr Raab, the sudden death from a heart attack story goes like this. First of all, a state of affairs develops in which there is some impairment of heart muscle oxygenation, due to a degree of 'hardening of the arteries' in which *plaques* of arteriosclerotic tissue reduce the lumen or bore of the coronary arteries. This lowers the amount of oxygen that is permeating the heart muscle. This so-called hypoxia, causes a lowering of the heart muscle's store of magnesium. At this stage a stress situation, be it psychological or physical, suddenly comes about which demands even more oxygen. This can further reduce the availability of magnesium in the heart. Under such circumstances the heart's action can lose its natural rhythm and start to beat in a disorderly and inefficient way. In some circumstances it can actually stop — and you are dead of a 'coronary', without even a thrombus to show for it at your autopsy!

The proof of the magnesium pudding

Some things are difficult to prove by means of direct experiment and the role that magnesium plays relative to cardiac health comes into this category. But in this case the 'inferential' proof is really quite convincing. One of the baffling things to doctors about sudden death from heart disease has always been the protective effect, from the heart's point of view, of living in a hard-water area. Residents of hard-water areas in the United States, such as Omaha and Nebraska, as well as people living in London, all have a much better chance of not dying suddenly from heart attack than those who are soft-water drinkers, like residents of the soft-water cities of the south-eastern United States and people in the soft-water areas around Glasgow in Scotland. It is highly significant that hard water contains more magnesium than does soft water. Hard-water drinkers therefore get a lot of their day-to-day magnesium from their tap water.

Doctors and scientists always like to have evidence from animal experimentation to support any new medical theories, and such evidence is not lacking here. In fact, it has been around for so long that we seem to have forgotten about it. Over forty years ago Dr O.M. Greenberg and some colleagues published a report in the *Journal of Biological Chemistry* which showed that if you reared rats on a magnesium-deficient diet they developed cardiac abnormalities. Other studies on experimental animals have shown that abnormal heart rhythms can be produced by short or long-term magnesium deficiency and that these are intensified by stress.

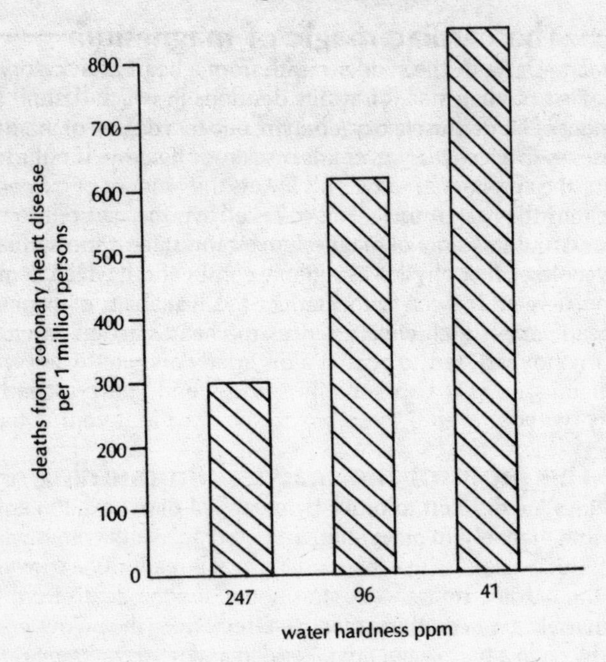

White men, 45 to 65 years old, in cities with softest, average and hardest water (Seelig and Heggtveit).

Figure 4
Deaths from Ischemic Heart Disease

Getting down to other cases

Apart from the serious sudden death type of emergencies that Dr Raab drew his colleagues' attention to, there is a wide spectrum of cardiac upset which makes many people's lives a misery — in fact, we met such a case in Chapter 1. These are the folk who suffer from what is called paroxysmal tachycardia, a highfalutin' name for a condition in which, for no known reason in most cases, the heart suddenly starts to race and you get bad palpitations.

Paroxysmal tachycardia patients may or may not have a degree of coronary artery disease. Normally, our heart beats at between sixty and ninety beats a minute and, although we can feel it at the wrist or in our neck, we normally do not notice it. Even if, under exertion or excitement, it increases well above 100 beats per minute, palpitation is not usually a worry.

The electrical circuitry of the heart is normally generated by a small collection of cells called the sino-atrial node and this is surrounded by cells of the autonomic nervous system. These are connected to the central nervous system and so when you are excited or frightened your brain can communicate directly with the sino-atrial node and 'tell' your heart to speed up — maybe you are going to be in for some action!

But there are other factors that are involved in the heart's smooth action too, and one of these is, as we have seen, magnesium. A patient described in Dr Mildred S. Seelig's book on magnesium deficiency was a 72-year-old woman who was admitted to hospital with a history of sudden onset of palpitations associated with fainting attacks. An ECG tracing showed the so-called premature ventricular systoles which meant that her heart was beating before it had a chance to fill properly with blood. But it went on beating rapidly and improperly. Luckily, her physicians were very much alive to the possibility of magnesium imbalance being the cause of her problems and when magnesium tests were completed, they demonstrated low levels of magnesium in her body. An intravenous injection of magnesium rapidly cured her symptoms. She was subsequently given supplementary magnesium and a year later had no heart symptoms. In retrospect, it was decided that this patient was a poor absorber of magnesium.

Alcohol and the heart

There is a wealth of undisputed medical evidence that chronic alcoholics develop bad symptoms of tachycardia when they are being 'dried out'. These 'alcohol tachycardias' are very often found to be associated with low blood magnesium. Less well documented are the experiences of normal social drinkers who find, as they advance into middle age, that if they go out to a cocktail party in the evening, or enjoy an evening meal with wine and liqueurs, they subsequently pay for their treat by a night's broken rest due to palpitations and tachycardia. What happens in such cases is that, possibly due to less than efficient magnesium absorption developing as they get older, their heart muscle is in a state of near magnesium depletion. As we have seen, it is an established fact that alcohol is what we call magnesuretic. In other words, it encourages the kidney to excrete more magnesium. When this occurs, the heart muscle reacts by becoming less than stable and the arrhythmia of magnesium deficiency is manifest. Regular magnesium supplementation usually clears this problem up.

Magnesium as medical treatment

In the late 1960s surgeons began to operate on the heart much more commonly for the repair of all sorts of cardiac abnormalities. To do so they used what is called a heart lung machine to 'take over' the function of the patient's heart during cardiac surgery, often for considerable periods. In some cases, once the surgery was over, the heart seemed to have trouble starting to beat regularly and efficiently again. Eventually it was found that such patients had low serum magnesium levels. It was decided to add extra magnesium to the pump system and the problem was rectified. Later, in the 1970s, when open-heart surgery was becoming commonplace, surgeons started to give intravenous doses of magnesium to their patients routinely, and the troublesome irregularity of the pulse of patients post-operatively became very rare indeed.

Magnesium has also been used quite extensively in the treatment of heart attacks, especially at a time when there was a vogue for the use of anti-coagulants in such cases. In one study, out of a series of 100 heart attack patients who were given magnesium therapy, only one died, compared with sixty out of 200 similar patients who were treated solely by anticoagulants.

What happens in low-magnesium heart cases?

Clearly, this is a complex business. In many cases 'heart' patients are treated with 'water pills' (diuretics). These slowly lead to magnesium depletion of the heart muscle which can subsequently become 'twitchy', as we have seen, and liable to malfunction.

But even without this stimulus when cardiac function is inadequate and angina is manifest, the heart's muscular metabolism is characterized by a loss of magnesium and potassium and a gain of sodium and water. Such a state of affairs predisposes the heart to irregular action and sudden and irregular failure.

A new study

The 'latest' in medical matters does not always mean the 'best', or even the wisest. But, of course, it is only natural for us to get a little bit excited at medical news items, especially when they are reported in prestigious newspapers such as *The Times*. Such a report, pertinent to magnesium's 'magic', appeared in this newspaper in July 1986 and it concerned experiments conducted at the Los Angeles County University of Southern California Medical Centre.

Using the method of atomic absorption spectrophotometry as an estimate of magnesium status in 103 patients admitted to a coronary care unit, researchers found that no less than 53 per cent had abnormally low magnesium status. Dr Robert Rude, one of the specialists involved in the study, emphasized the link between cardiac arrhythmias and magnesium deficiency, and divulged that magnesium-deficient heart patients at the University Medical School's Cardiology Division are now being given magnesium supplements.

CHAPTER 5

Magnesium and Women's Problems

In Chapter 1 we met Rebecca, a 35-year-old mother of two who suffered with the pre-menstrual syndrome (PMS; also called pre-menstrual tension or PMT). Her problem was so bad that her husband began to feel she was undergoing a personality change. In fact, not to put too fine a point on it, he wondered if she was going crazy. One thing was certain — she tended to drive *him* crazy regularly for a week, every month! He had seriously considered leaving her, and may well have done so had it not been for one thing. The other three weeks of the month she was the sunny, charming and lovely woman he had fallen in love with and married!

As it happens, I could have introduced you to a score of other PMS sufferers. Here are some examples. Alice found that her PMS symptoms first started when her youngest child, Michael, first went off to school all day. Suddenly she felt 'almost redundant', she said, and although she had looked forward to the time when the 'children were off her hands more', as it happened PMS arrived to make her life a misery instead. Andrea's PMS seemed to be stress related in another way. The whole thing started when her husband Bill's firm moved offices to the North of England. It did not seem reasonable for the children's schooling to be badly disrupted and so Bill took 'digs' near his firm four nights a week and came home at weekends. This made Andrea feel even worse. Jill's case was different again. She had had three children. She could not take the pill and birth control had always been a bit of a nightmare for her (her last baby had not been planned). So she was sterilized. With sterilization came PMS.

Sally had taken the pill successfully for years, but at 35 her doctor had said 'enough is enough' and had fitted her up with an interuterine contraceptive device. It had proved to be very efficient, but the next month, after stopping the pill, PMS had started bothering her.

Tracy's case was different again. She was absolutely delighted with her first baby and had breast-fed her for nine months. Then she wanted to take a part-time job and so had weaned little Sophie who did very well on her artificial feeds, but Tracy suddenly developed PMS.

Young Zoe had had what you might call an anxious adolescence. She had been nearly 15 before her breasts started to develop very much, and was 17 before her periods started. Everyone had sighed a sigh of relief, but before long Zoe was back seeing her doctor with PMS of the type that late developers often seem to get.

Patricia was another special problem. She was 27 and had been trying to get pregnant for three years. The specialist at the infertility clinic put her on a 'fertility pill' (clomiphene) but she could not stick it due to the really nasty PMS that developed.

Marjorie was really looking forward to the change of life! Her periods had been heavy for the last four years and her doctor had on several occasions suggested an operation — a hysterectomy. But she had decided, at 48, to 'hang on'! Surely soon the problem would be cured by Nature. But Nature was in no hurry! She decided to visit yet another problem on poor Marjorie — a really devastating episode of PMS! Like so many women she had heard about it from her friends and neighbours, and had read about it in women's magazines. But now it was happening to her she decided that there was one thing she wanted to know from her doctor. What exactly is PMS and what causes it?

PMS — a doctor's dilemma

The pre-menstrual syndrome is a unique disease in a way, for it was more or less discovered by a French criminologist over a hundred years ago when it was noted that 63 per cent of all women apprehended for shoplifting were in a pre-menstrual phase of their monthly cycle. But it was not until 1931 that doctors became interested in the condition, when an American gynaecologist called Robert Frank described a series of fifteen women who suffered from a syndrome (syndrome means a collection of symptoms) that involved irritability, anxiety, depression and a feeling of being blown up or bloated in the days before they menstruated or in the first four days of the menstrual flow.

Today we have refined that excellent description only a little and define

PMS as the occurrence of a variety of apparently unrelated symptoms that are seen only in the second half of the menstrual cycle and which disappear during the early days of the period. Two types of PMS seem to occur. About 90 per cent of women experience a mild form lasting three or four days. Usually they shrug it off as one of the curses of Eve and rarely seek much in the way of relief.

Then there are the more severe cases. They exhibit quite different symptoms, such as marked tension, upsetting degrees of anxiety, fluid retention and breast discomfort. What is more, these unpleasant symptoms may last from seven to fourteen days before the period starts. Distress is often severe enough to disrupt family and working relationships. Normal activities are curtailed. In other words, here we have a real illness that is as deserving of treatment as any other medical condition.

Arguments about the fundamental cause of PMS have preoccupied research workers for years now. Because of the cyclical nature of the problem clinical trials have been difficult to design and interpret. Many of the classic symptoms of PMS are difficult to quantify. For instance, how do you measure a feeling of being bloated or irritable? It has even been found very difficult to conduct the so-called 'double blind' trials in which nobody (that is neither doctor nor patient) knows who is receiving a drug under test for evaluation and who is receiving a dummy placebo tablet.

Theories galore

And so the whole subject of PMS is rich in theories, but it tends to be poor as far as hard scientific fact is concerned. An early theory that PMS was largely a psychological condition has pretty well been abandoned now. This is not because of any lack of psychological symptoms in the disease, but because so often the psychological symptoms are cleared up by treatments of a non-psychotrophic nature! Be this as it may, PMS can certainly trigger off psychological problems which is rather different. For instance, women who are known to suffer episodes of psychiatric illness are much more likely to relapse and be admitted into hospital because of these conditions during the pre-menstrual days of their lives than at other times. It also seems true that women who are psychiatrically ill suffer PMS more severely than their fitter sisters.

The most fashionable theory for PMS today is that the disease is due to some imbalance between the two sex hormones, progesterone and oestrogen, which, by means of a very neat biological ebb and flow system, control and regulate the whole process of ovulation (egg cell release) and menstruation. The previously held 'hormone theory' for PMS (that it was simply due to lack of one hormone — progesterone) is no longer tenable because many investigations of PMS victims have disclosed normal progesterone levels in sufferers. Nevertheless, giving extra progesterone does help some PMS victims, probably because this has some central effect on the centres that regulate the whole intricate system of female sex biochemistry.

A certain amount of research has implicated another female hormone called prolactin in the PMS. Prolactin is sometimes referred to as a stress hormone. As the name suggests, this hormone certainly stimulates the breast. It also gives rise to the retention of sodium and potassium, as well as water, by the body. Unfortunately, prolactin is a very difficult hormone to measure — even approaching a patient with a hypodermic syringe will cause the body's prolactin levels to rise making accurate assessments tricky.

Magnesium and PMS

One of the most interesting studies on the subject of PMS has recently been carried out by a group of doctors from the Royal Sussex County Hospital in Brighton. This was stimulated by research carried out in the United States by Dr G E Abraham who had first suggested that many of the diverse symptoms of PMS might be due to magnesium deficiency.

Interestingly enough, a survey carried out which involved the women members of fifty colleges in the United States showed that the food which they ate provided very definitely sub-optimal levels of magnesium. As we have seen earlier, women need at least 300mg of magnesium per day to maintain themselves in magnesium balance and yet the magnesium taken by the average female studied in this survey averaged out at only 204mg per day. This does not seem to be an isolated example of low magnesium intake in women. A doctor practising in Tennessee also went to the trouble of carrying out a seven-day metabolic study of the magnesium status of ten women in his practice. Their mean daily magnesium intakes were only 60 per cent of the recommended intake.

Dr Mildred Seelig, magnesium expert *extraordinaire* who works at the Goldwater Memorial Hospital in New York, feels that the low magnesium status of women is liable to be eroded even more by more recent changes in eating patterns. Women are increasingly involved in 'eating themselves thin' by increasing the amount of fibre, and phytate, in their diets. They do this by increasing the quantity of such foods as brown and wholemeal bread, brown rice, and oatmeal in their everyday eating. This is fair enough from the slimming point of view but these and other fibre-rich foods tend to relentlessly reduce magnesium absorption.

In the Brighton study cited previously, 105 patients who were PMS sufferers were recruited from two sources — a private practice and a well-woman clinic run by a local health authority. All women were shown to be suffering from PMS by means of a diagnosis questionnaire developed by Dr R H Moos and first reported in the *American Journal of Obstetrics and Gynaecology*.

A sample of blood was taken from each patient. Anybody taking vitamin B_6 or magnesium was excluded from the survey. The method of determining the magnesium status of the group studied was by estimating the amount of magnesium that was present in both the plasma and the red blood corpuscles by the method of atomic absorption spectroscopy. The mean erythrocyte magnesium of the PMS patients was significantly lower than in controls ($P<.001$). There was no significant difference in the plasma magnesium between the PMS groups and controls. This was to be expected because plasma magnesium remains remarkably constant even in cases in which patients' magnesium is in negative balance because, as mentioned in Chapter 4, magnesium is held *inside* cells rather than in body fluids.

In the laboratory in which the red cell magnesium tests were carried out they have made several studies of erythrocyte magnesium in a variety of disorders and have discovered that PMS is the *only* common condition in which red cell magnesium is lower than their established normal range.

The logical management of PMS

I would hazard a guess that before very long any scientific appraisal of treatment methods of PMS will be preceded by an estimation of red cell magnesium so that research workers will be absolutely sure that they are involved in the treatment of PMS. For the time being, however, the most likely way of managing the syndrome is to make the diagnosis *reasonably certain* by the use of a menstrual diary chart. There are several of these available and the *proforma* illustrated is one that works

Figure 5

How to Complete a Menstrual Chart

Fill in a menstrual chart, using the appropriate letters from the key below on the days symptoms trouble you.

days of month

month	1	2	3	4	5	6	7	8	9	10	11	12	13	14	15	16	17	18	19	20	21	22	23	24	25	26	27	28	29	30	31
1st month																															
2nd month																															
3rd month																															

key to symptoms:

- M — menstruation
- T — tension and irritability
- D — depression
- B — bloated feeling
- P — pain, headache or backache
- F — fatigue

Figure 6

This chart suggests a diagnosis of PMS

days of month

month	1	2	3	4	5	6	7	8	9	10	11	12	13	14	15	16	17	18	19	20	21	22	23	24	25	26	27	28	29	30	31
1st month March																		T	T		T	T	T F	T B F P	T B F P	M	M	M	M		
2nd month April																T	T	T F	T B F	T B F	T B F	M F	M	M							
3rd month May.															T	T F	T B F	T B F	T B F	T B F	M	M	M	M							

reasonably well in practice. Charts can be used as a 'do it yourself exercise' to make sure that a woman is suffering from typical PMS and not another sort of psychological or menstrual problem.

In most cases it is a good idea to discuss the whole problem with a doctor at an early stage as this gives the woman in question a chance to voice any fears that she may be harbouring about herself and how she is feeling. Some women feel so terrible with their PMS that they are sure they are going mad or suffering from something awful like a brain tumour.

The very fact that it is possible to talk to someone about a problem is itself very therapeutic. The fact that the chart shows, as it will, how the various symptoms come and go with the ebb and flow of the periods is often tremendously reassuring too.

First treatments — gentle treatments

There's an old saying that you don't need a sledge hammer to crack a nut and this is particularly apt as far as the management of pre-menstrual tension is concerned. So if a simple and straightforward discussion with a PMS expert, be it doctor, social worker or fellow sufferer, does not prove helpful then it is time to try some form of suitable therapy. Traditionally, vitamin products are usually tried first. Vitamin B_6 is often useful and has proved beneficial both in uncontrolled and to some extent in controlled trials. There is one school of thought which suggests that there are several forms of PMS rather than one stereotyped condition. In other words, in one form fluid retention is the predominant symptom. Another form ruins a victim's life because of severe breast discomfort. In yet another form, it is 'psychological' symptoms that dominate the picture — for example anxiety, depression, mood swing, aggression, anti-social behaviour, recklessness and so on. These are the cases in which vitamin B_6 seems to work best in doses of 100mg daily from day fourteen to the onset of the period.

Magnesium therapy has really not been around long enough to obtain a firm niche in any authoritative way but clinical trials are underway and results are already promising. Magnesium also has the advantage of being free from side-effects and is relatively cheap.

The third line of gentle approach to the problems of PMS is best looked upon as prostaglandin manipulation. Prostaglandins are relatively recently discovered local tissue hormones. The first hormones discovered by medical science were what you might call long-range chemical messengers. For instance, the thyroid hormone called thyroxine is produced in the thyroid gland in the neck and travels via the bloodstream all over the body, regulating the rate at which the body uses foodstuffs by converting them into energy.

The prostaglandins are quite different. They are produced *locally* in tissues and bring about their effects *locally* too. Doctors have been influencing various disease processes in the body by using drugs that alter prostaglandin action for some years now. The best known of these new therapeutic bombshells has been the production of prostaglandin inhibitors (like Nurofen) in the management of rheumatism. The other side of this curious prostaglandin therapeutic coin is prostaglandin enhancement and this brings us back into the field of basic nutrients with a bang. Linoleic and gammalinolenic acid encourage certain prostaglandin syntheses. In fact, gammalinolenic acid is the precursor of prostaglandin PGE, in the body, a substance that has complex interactions with prolactin and other hormones at a tissue level.

Many symptoms experienced by PMS sufferers can be related to the wide spectrum of actions controlled by prostaglandins. What is more, peak levels of prostaglandin production tend to occur cyclically in the body — as indeed does PMS.

Strangely perhaps there is a link forged between magnesium and prostaglandins. Magnesium is one of the agents that the body uses to convert linoleic acid into gammalinolenic acid (sometimes referred to as GLA). GLA is subsequently converted into prostaglandins. It appears that vitamin B_6 is also involved in the conversion of GLA into prostaglandin, as are vitamin C and zinc.

More 'medical' management of PMS

Wise physicians will always try the gentle remedies described above before advising their patients to seek out more pharmacological remedies. When they do so, they often practise the following routine, although individual doctors may well subscribe to individual treatments.

For women who are not anxious to conceive and for whom general medical considerations make it possible, a trial on a standard oestrogen/progesterone contraceptive pill is often an early medical reaction to the more symptom-resistant type of PMS sufferer. Women who smoke and are over 35 are not, generally speaking, considered suitable for such treatment, however. Patients who are unsuitable for Pill-type therapy in PMS, or those who in addition to PMS also suffer from painful periods, often react better to treatment with dydrogesterone or progesterone. For the type of PMS in which breast symptoms dominate the condition, bromocriptine or danazol are often prescribed.

For some women troublesome bloating and general symptoms of severe fluid retention are the major problem of their PMS. In cases like this, diuretics used to be widely prescribed and although these will often bring about a temporary and sometimes embarrassing relief of bloating symptoms, in which the poor victim will hardly be able to leave the lavatory, other symptoms tend to get worse possibly due to the fact that diuretics tend to make the body lose more magnesium. In cases like this the drug spironolactone is sometimes effective.

CHAPTER 6

Magnesium and Your Blood Pressure

People with high blood pressure generally feel just the same as people with normal blood pressure, so how do you know if you have it or not? You may find out as a result of a medical examination, maybe for a life insurance or to enter a pension scheme, or perhaps because you fall ill with something quite unrelated to blood pressure — say, with appendictis — and somebody takes your blood pressure as a routine procedure.

Most people have only the vaguest idea as to what constitutes high, low, or normal blood pressure and until recently even doctors have had trouble deciding on these tricky points. But as a result of a lot of painstaking research some of the 'fog' that has obscured this whole subject is lifting.

First of all, it is necessary to realize that we all need our blood pressure! When the blood pressure falls profoundly we become unconscious. This is what happens when the guardsman passes out on parade, or the medical student faints during his first operation session at the hospital. It is also what makes children collapse at school assembly. Understanding why these things happen tells us a little about the practical side of normal blood pressure. The guardsman passes out because he has been standing for a long while in one position. You may not realize it but in a way your legs are your 'second heart' pumping back blood to your 'number one' heart as you move them. The man or woman 'on parade' or the child standing still in school assembly does not have the advantage of his blood being briskly returned to the heart from the general circulation and so the blood pressure falls and consciousness lapses.

The medical student or nurse who faints in theatre does so because of a strange sort of psychological panic reaction. It is as if the brain says

'I don't think I can take any more of this traumatic excitement, I must escape!' But escape in such circumstances is not all that easy and we react by unconsciously allowing all the blood vessels in the body to lose their normal 'tone'. The blood then rapidly 'pools' in the gastrointestinal circulation not leaving enough blood pressure in the brain to maintain consciousness. As the victim sinks to the floor, however, blood pressure is restored to the central circulatory system, with the help of gravity, and the victim recovers. A similar state of affairs was evident in Victorian times when ladies 'swooned away' when difficult psychological or emotional situations developed. (In this case two other things made them prone to swooning as well. Iron deficiency anaemia was commoner in those days then it is now and anaemic blood carries less oxygen than it might and so makes the victim liable to faint. Also the commonly popular 'tight lacing' of Victorian times restricted blood flow and further reduced effective blood pressure regulation.)

Normal blood pressure

The blood pressure is always expressed as two numbers and 130/80 might reasonably be looked upon as normal blood pressure. The first figure (130) is the blood pressure inside the large arteries in the body at the moment that the heart is in contraction. The heart has at this moment just pushed a 'heart full' of blood out into the arterial circulation. The second figure (80) is the pressure inside the large arteries when the heart is relaxing. Both of these pressures are measured in units and these units are millimetres of mercury. Think of a U-shaped glass tube several centimetres high propped up against a wall and containing some mercury. If you just left it there the liquid mercury would be level in both arms of the U. If you put one of the ends of the U in your mouth and blew hard, the mercury would move up in the arm of the U that you were not blowing into. If you measured how far up the tube you could blow the mercury compared with the resting non-blowing mark (in centimetres) you would have measured the power of your blowing force.

For years doctors have more or less agreed that the blood pressure rises normally with age. A rough and ready 'guesstimate' of systolic (heart contracting) blood pressure is your age plus 100. So a 50-year-old man would have a systolic blood pressure of 150. The diastolic resting figure is much more constant and most doctors agree it is much more important than the systolic pressure. Readings over 90 are always looked upon with suspicion by doctors as being possibly worrying.

So a 50-year-old man has a normal blood pressure of 150/90, you might say. But not always! It can go up and down with excitement, stress, exercise and, strangely, it will fluctuate week by week, running a little high this week, a little low next week. All this boils down to the fact that the wise physician never commits himself to a diagnosis of high blood pressure on one blood pressure reading — or even two or three. He will want to assess this on several occasions, with his patient sitting and standing, in the morning and the evening, before he decides to diagnose hypertension and, if necessary, start treatment.

The treatment of hypertension

The diagnosis and treatment of high blood pressure is in no way a 'do-it-yourself' type of health exercise. It must be left entirely to a physician to decide (a) who has high blood pressure and (b) whose high blood pressure warrants medical or any other sort of high blood pressure management (for example, drug treatment, weight reduction treatment or relaxation therapy). I have dealt with this interesting aspect of this very important subject in *The Ten Day Relaxation Plan* (Piatkus, 1984).

There is, however, a way in which the magic of magnesium has a bearing on this whole subject. It may mean that by simple manipulation of our magnesium intake we can, over the years, make the likelihood of suffering from the *sequelae* of hypertension like coronary disease and stroke illness less likely. For there is a certain amount of evidence that magnesium keeps our arteries healthy.

The artery in hypertension

One of the ways that a doctor assesses his patient as far as artery health is concerned is simply to feel the pulse. He is not only counting the pulse rate as he does this. He is also deciding if the artery feels elastic and flexible or hard and rather rigid. During a medical examination the physician will often assess the health of other arteries too. The one that runs down just in front of the ankle is a favourite one, for this artery often becomes hardened and rigid in arteriosclerosis and other forms of artery disease. Finally, the good doctor will always take a *look* at your arteries — he will really look at them in the 'raw state' (not covered with skin). He can do this by examining the back of your eye as they track over your retina with an ophthalmoscope.

The reason for all this interest in artery health is that high blood pressure and artery disease go hand in hand. We do not know, for sure, which comes first, high blood pressure or diseased arteries. Perhaps the first change to occur before the blood pressure rises significantly is for the wall of the artery to become hard and rigid due to cholesterol and calcium being laid down in it. This makes the heart try to maintain good circulation by pushing up the blood pressure. Or maybe the blood pressure goes up for some reason and as a result the arteries get cholesterol and calcium deposited in them. A chicken and egg situation *par excellence*, you will agree.

But to get back to minerals and magnesium, it has been repeatedly noticed by research workers in the field of hypertension that *before* arteries get really 'sick' due to fatty infiltration and cholesterol, prodromal changes can sometimes be seen in them. This can even happen in infants and children. Usually these changes occur in small and medium-sized arteries, in which the elastic tissues in these vessels undergo a sort of special degenerative change. It has been noted repeatedly by research workers that these changes in infantile arteries have a quite striking resemblance to the changes that occur in animals' arteries when such creatures have been raised on magnesium-depleted but otherwise normal diets.

This research finding has reminded doctors that children and infants do develop, and even die, of coronary disease, though this is quite rare. It is not outside the bounds of possibility that magnesium deficiency may be implicated in the very earliest stages of artery disease for the following reason. Children raised on formula feeds take in with their food a substantial phosphate load, together with perhaps rather more vitamin D than Nature intended. This is due to differences in composition between human milk and formula feed milk (cow's milk). Both phosphate and vitamin D tend to lower tissue magnesium levels.

Animal experiments have repeatedly confirmed that a similar state of affairs can be duplicated in the laboratory. Artery changes can be produced 'to order' in rats by feeding them on a low magnesium high calcium diet, and the addition of a high proportion of fat to the diet increases the degree of arteriosclerosis. In fact, unkind but dedicated research workers have during the 1980s devised an experimental diet that so damages the arteries of the rats, dogs and cockerels that were fed on it, that 80-90 per cent of these animals died of coronary disease. The diet was high in fat, cholesterol, vitamin D, phosphate and protein, and low in magnesium, potassium and chloride. It was nicknamed the *cardiovasopathic* diet and it is remarkably close in many ways to the sort of food we tend to favour in our western society for everyday consumption.

Magnesium and blood pressure

Is magnesium therefore a shy Cinderalla once again, working away quietly in the kitchen, protecting the health of our arteries if we will let her? There is a fair amount of evidence available today that makes such an outlook seem very reasonable. We know, for instance, that if we suddenly reduce the body's magnesium supply — for example, as a result of replacing lost body fluids that occur in severe gastroenteritis with magnesium depleted fluids — there is a sudden rise in the blood pressure. Conversely, if large doses of magnesium are given to patients with a severe degree of hypertension then hypotension (low blood pressure) develops quite rapidly.

It seems more and more likely that high blood pressure gradually predisposes us to the disease we call arteriosclerosis by damaging the cells that line the insides of our arteries, and that this damage allows cholesterol to become deposited in the artery wall, especially at places in arteries where swirling or eddying of the blood stream occurs. These twists and turns of the arteries may well be pressure-induced. Such changes are more frequently seen in magnesium-depleted animals than in those who have been fed a magnesium-adequate diet. Magnesium-deficient animals often develop hypertension, and subsequently artery and cardiac damage. Significantly too they also have high blood fat levels and low blood calcium levels. All in all, it seems difficult to resist the conclusion that magnesium keeps blood vessels healthy. Healthy arteries prevent heart attack and stroke illness, and many other ills that flesh is heir to.

Admittedly, we do not have direct and convincing medical evidence that is totally supported by controlled and 'double blind' clinical trials, to demonstrate that magnesium is a valuable long-term prophylactic as far as providing us with a stable and normal blood pressure. Due to very difficult practical difficulties, it would seem more than likely that we shall never have this sort of firm supporting evidence. And so we have to look at what we do have and draw our own conclusions. On the whole, a mass of animal experiments support the hypothesis that magnesium depletion leads animals of many species along a path that in turn leads to premature arteriosclerosis and higher than normal blood pressure.

From childhood onwards, people in the western world seem firmly committed to living on a diet that has been shown to be cardiovasopathic at least to rats, dogs and cockerels and is, as stated previously, capable of producing coronary thrombosis in 80-90% of the animals that consume it. Moving ourselves away from the cardiovasopathic diet (now becoming known as CVD) involves eating less fat, and less vitamin D, sodium, phosphate and protein — dietary options that are not always easy to take. But such a diet can be improved in one quite simple way for the CVP is low in magnesium and potassium. In other words, magnesium and potassium supplementation on a long-term basis can help prevent some of the dangers of improper eating.

Finally, there is an item of medical management relevant to this whole field that calls for a fairly urgent rethink on the part of doctors everywhere. Diuretics are still sometimes used to lower high blood pressure, although not as often as they once were. These are sometimes called 'water pills' by those that take them, due to the fact that shortly after their use the patient has to spend a considerable portion of the day passing his or her water. But, as we have already seen, diuretics are liable to cause considerable magnesium loss through the kidneys. Is it still reasonable to use a drug that we now know reduces tissue magnesium when it appears that low tissue magnesium makes us all more vulnerable to coronary thrombosis, heartbeat irregularity and high blood pressure? I for one would say that diuretic antihypertensive therapy needs an urgent and strict clinical reappraisal and I would be loath to submit to it personally.

CHAPTER 7
Magnesium and the Health of Your Kidneys

A problem with kidney stone is something that we could all do without. Nevertheless, it is something that could strike you down at any time. Kidney stone is a common and potentially serious disease but a considerable degree of prevention does seem to be possible these days. The vast majority of kidney stones contain calcium, mostly in the form of phosphates or oxalates, and sometimes even a mixture of the two. Rarer types of kidney stone are made of uric acid or cystine.

The food we eat contains all the elements of kidney stones and so in a way they all come from food. The calcium in them comes mostly from milk and dairy produce, the oxalates from rhubarb, spinach and asparagus. Phosphates are present in most of the foods we eat every day and so is uric acid and cystine. But of course, although we all eat these things, luckily only a minority of us get kidney stone problems. Why this is so has taxed the deductive power of doctors for many a year. Now at last there appears to be more than a hint of a better understanding of the problem.

In their search for something in the nature of a 'trigger factor' as far as kidney stones are concerned, all sorts of interesting things have been suggested. There are a couple of tiny glands which sit on top of our thyroid glands (called the parathyroid glands) and sometimes these become overactive. One of the problems which this *hyperparathyroidism* produces is kidney stones. But only very rarely is there parathyroid disease in stone victims. Certain bone diseases, such as rickets, osteitis, and osteomalacia, are also associated with kidney stones. But again most stone sufferers are free of these diseases.

Finally, kidney infection or anything that interferes with the normal outflow of urine predisposes towards stone formation and a very few of the hundreds of stone sufferers will be faced with such problems.

In the large majority of cases of kidney stone, none of the predisposing causes of stone formation seems relevant and the stones 'just come'. And when they do come they signal their arrival in various unpleasant ways.

Is it the stone, doctor?

Exactly what happens to the kidney stone victim is pretty variable. Sometimes the stone may be passed out from the kidney into the ureter (which connects the kidney to the bladder). When this happens, the victim usually experiences a nasty attack of pain that starts in the loin or the small of the back and then radiates down into the groin. With luck, the stone finds its way into the bladder and may then stay in the bladder or be voided (or 'passed') in the urine.

Some kidney stones announce their presence in other ways. For instance, they cause local damage to the urinary tract and blood appears in the urine. If they obstruct the flow of urine, urinary infection may follow and the victim is liable to run a temperature and have pain on passing infected urine.

Doctors are always most anxious to preserve kidney health, for kidney damage can be a very serious business indeed and so they enthusiastically investigate the potential stone victims using various x-rays and laboratory tests. Sometimes kidney stones need to be removed, which means surgery in most cases, although there is a new method of treatment that 'persuades' kidney and bladder stones to 'explode' into small 'passable' pieces due to ultrasonic shock waves.

'Cutting for stone', the surgical removal of stones, was one of the earliest surgical operations to become 'fashionable' in past centuries. Many famous surgeons made their name for the accuracy and speed by which they 'cut for stone'. The advantage of speed will be appreciated when we remember that no anaesthetic, other than a large dose of alcohol, was available in those days. However, victims gladly submitted to surgery, so nasty is the pain of renal colic.

Diet and stone formation

Way back in the 1950s it was realized that all of our kidneys contain microscopic deposits of calcium in them, which are believed to be normally removed by the kidney lymphatics. When the kidney stones form it is thought that these mini-stones, or *microliths*, as they are called, suddenly decide for some reason to build up and form plaques or patches. An early experimental worker by the name of Greta Hammarsten found that if she fed rats on a very low calcium diet, the animals mobilized the calcium they needed from the bones. But somehow, once this process was initiated there was no easy way of stopping it and an excess of calcium started to appear in the urine. She also experimented with low magnesium diets. These resulted in a low level of magnesium in the urine and this tended to make oxalates in the urine become insoluble and calcium oxalate stones started to form quite easily in her experimental diet. She even worked out a 'stone producing diet'. It was low in calcium, low in magnesium and low in vitamins A and D. It also tended to be high in oxalates.

Back to people

Although animal experiments are always interesting (more to the experimental scientists than to the animals, naturally), a link between magnesium and kidney stones in humans really started to be forged over twenty-five years ago when an editorial in Britain's prestigous medical journal, the *British Medical Journal*, drew its readers' attention to the fact that kidney stones were on the increase in Finland, Central Europe, Sweden and Japan and pointed out that these areas of high calcium/oxalate kidney stones incidence coincided with areas in which there was a rising incidence of cardiovascular disease.

Later on in 1975 a map of the United States was produced which pointed out dramatically that in the areas where the water was softest in the USA, the incidence of kidney stones in the population was highest. We have already seen how hard and soft water affect heart conditions (page 41). It seems that similar conclusions can be drawn in the case of kidney stone. The medical geographers were soon to draw a 'kidney stone belt'. It included South Carolina and the south-eastern states, which are soft-water areas, while the mid-west and south-west earned themselves the epithet of being 'good kidney country'. The more American physicians delved into this somewhat astonishing relationship, the more definite it became. Nebraska, a state well known for the cardiac health of its inhabitants (it enjoys the lowest incidence of sudden death from coronary disease in the USA), also had the lowest frequency of urinary stones. Finally, Dr R R Landes, American expert on the subject of kidney stones, went on record in 1976 as saying that a statistically significant relationship

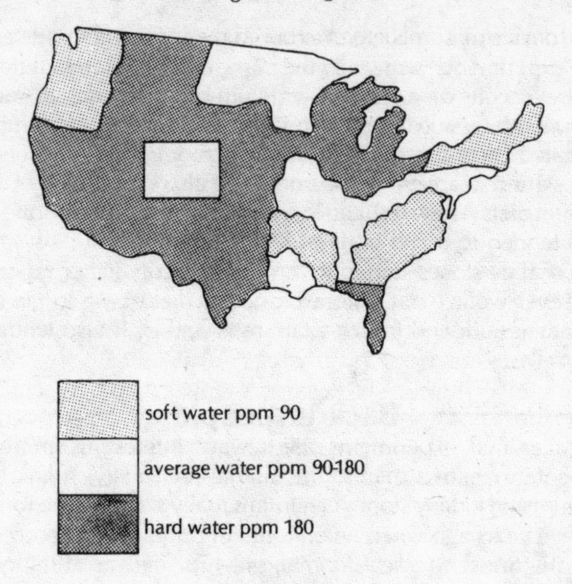

Figure 7
United States Water Hardness Map
1963 (Landes)

between kidney stones and the hardness of the water had been finally demonstrated.

Some medical geographers have drawn conclusions relevant to this whole problem that explain seemingly strange variations in kidney stone incidence on a time-scale basis. Shortly after the First World War, high incidence of kidney stone occurred particularly in Europe. This coincided with the use of artificial fertilizers in farming that were low in magnesium which in turn brought about magnesium-deficient crops.

–More news on the production of kidney stones–

Quite slowly our medical profession had noted Dr Hammarsten's experiments on rats in the late 1920s which demonstrated that an increased amount of magnesium in the diet increased the solubility of calcium oxalate in the urine. Eventually two doctors working at America's famous Johns Hopkins Hospital showed that if magnesium was added to the urine of kidney stone victims, it lost its facility to produce crystals if it was cultured with live connective tissue, a procedure thought to mimic natural stone formation in the body. Experiments on patients, rather than in test tubes, followed. Magnesium was given to a group of patients who repeatedly passed urine containing crystals of calcium oxalate — in other words, patients at high risk of developing kidney stones. The condition rapidly cleared.

Two other doctors studied the effect of giving extra magnesium to urinary stone victims. It was shown to prevent both high urinary calcium excretion and new stone formation. In one such patient, his illness soon relapsed when the extra magnesium he had been prescribed was withdrawn from his diet. Dr E L Prein, writing in the *Journal of the American Medical Association* in 1965, reported that he had given vitamin B_6 and magnesium supplementation to a series of fifty patients suffering from recurrent kidney stone. Most showed a reduction in the incidence of kidney stones.

Subsequently, other quite large numbers of stone patients have had favourable results as a result of long-term treatment of their calcium oxalate kidney stone disease with magnesium.

Today the numbers of recurrent stone sufferers whose nasty disease has been virtually abolished by the therapeutic use of magnesium runs into thousands. Strangely, perhaps, advice from the medical profession on magnesium supplementation, especially in soft-water areas, seems to be largely lacking.

It seems highly likely that, in the interests of keeping well away from the terror of kidney stones, particular attention should be paid to the maintenance of an adequate magnesium intake. Those who believe that this form of 'kidney insurance' is worth the 'premium' of a daily magnesium supplement will be doubly fortunate if they opt for a supplement that also includes vitamin B_6 as this too is earning itself an enviable reputation in the field of renal health.

CHAPTER 8

Magnesium and Strong Bones

Ethel Green was a most helpful old lady to have as a neighbour. Although well into her middle seventies she kept her house spotless, did all her own shopping and was only too pleased to 'take the key' of anybody's house in the neighbourhood when they went off on holiday. She could also be seen walking each day, usually with two or three dogs, for she loved animals and the countryside. Being out in all weathers was, she said, half the fun of being alive!

Then one day somebody passing her house heard a plaintive little voice crying for help and Ethel was found in her tiny back garden lying among her summer vegetables. An ambulance was called and she was taken to hospital. X-rays disclosed that she had fractured her femur (thigh bone) just at the top or neck of the bone, where it fits into the hip joint. Some time later, an orthopaedic surgeon operated and a steel pin was inserted to join the fractured parts of her femur together. It seemed highly likely that this old lady would walk again, but maybe without three dogs on leashes!

One thing really bothered Ethel Green over this whole episode. All sorts of people who came to visit her in hospital remonstrated with her about doing 'silly things', things in which she was liable to fall and 'break a bone at her age'. Everybody knows that elderly people's bones get brittle and the 'slightest fall can mean a fracture'. But, as Ethel repeatedly said, she did not fall, trip or stumble! One minute she was dead-heading her roses and the next minute she was on the ground. Her leg just gave way. This all happened a couple of years ago. Maybe today somebody at the hospital would *believe* a story like Ethel Green's for slowly doctors and medical personnel generally are beginning to believe the 'little old ladies' who suddenly find themselves on the floor with a fractured femur.

In many such cases they have not tripped or fallen. What has happened is that their thigh bone has broken — and *then they have fallen*. A disease of excessive brittleness of the bones called *osteoporosis* has been present for some time. It can make bones so brittle and fragile that they can be broken by the forces of gravity or muscular action alone. Of course, a minor stumble or fall can fracture them too. But in a number of cases the fracture occurs before the 'accident', not as a result of it, and so we must be more careful before we 'tick off' such elderly folk for taking so-called foolish risks.

What exactly is osteoporosis?

This is a good question that is only just being answered by physicians and rheumatologists the world over. First of all, osteoporosis is an age-related disorder of bone. It is characterized by decreased bone mass and by an increased susceptibility to fractures in the absence of any other obvious cause. It is very common. The Office of Medical Applications of Research in Bethesda, Maryland in the United States, is on record as stating that in the USA there are 15-20 million sufferers and the disorder causes about 1.3 million fractures each year. Mostly they are in folk over 45. If you are lucky enough to reach the age of 90, there is a 32 per cent chance of you suffering from a hip fracture if you are a woman and a 12 per cent chance if you are a man.

In most cases osteoporosis is undiagnosed until complications occur. A couple of years ago the *Journal of the American Medical Association* estimated that osteoporosis costs the USA 3.8 billion dollars annually. Unfortunately, in the UK to do such 'health sums' is not considered to be a suitable exercise, particularly in the face of governmental economies in National Health Service expenditure where 'expensive' diseases are bad news for government. But it seems likely that an enormous amount of suffering and expense occurs from osteoporosis in every country in the world, and if we really look for it we find it.

To understand a little better about osteoporosis we have to understand a little more about bone. Bone is basically a gelatinous tissue that is intimately impregnated with minerals, notably calcium and phosphorus but also with magnesium. It is a mistake to think of bone as an inanimate sort of tissue. Admittedly it looks pretty dead and dusty when you see a skeleton hanging up in a first-aid lecture, but bone in fact undergoes a continuous remodelling (turnover) all through life and is very much alive! Cells called osteoclasts constantly absorb bone while other cells called osteoblasts reform bone. Normally bone resorption and formation are linked closely in time and all sorts of factors, including mechanical and electrical forces and hormones, influence the remodelling or 'turnover' process.

We all reach our peak bone mass at the age of about 35 but sex, race, nutritional profile, exercise and overall health all influence peak bone massing to a variable extent. In men the total mass of bone at its peak is 30 per cent more than it is in women. There is an interesting racial variation too for bone mass is about 10 per cent higher in black people than it is in white races.

The downhill road

Life is said to begin at 40, but I am afraid that the same cannot be said of bone mass. At the age of 35, after reaching its peak, bone mass starts to decline due to an imbalance of the modelling process, and bones start to lose both their mineral and their gelatinous matrix. But they still retain their basic organization and function quite well if given the chance. In women the menopause, or change of life, brings a sort of crisis in bone remodelling which lasts for between three and seven years, during which a rapid decrease of bone mass occurs. Treading the downhill path is of course inevitable in a way and is reflected in what happens to us as we progress through life. Thus we see women sustaining more fractures than do men — and whites more fractures than blacks.

Bones seem to differ as far as osteoporosis and the fractures associated with it are concerned. Fractures in the spinal bones (vertebrae) occur most frequently in women aged between 55 and 75, while hip fractures occur most commonly in older folk of either sex.

To some extent the progression down the road of disability seems to have been accepted by the medical profession as being inevitable once an osteoporosis fracture has occurred. Sadly, most patients with hip fractures fail to recover their previous degree of activity, and nearly 20 per cent die during the following year. Fractures of the spinal bones which develop while 'doing what comes naturally' — that is, bending down, lifting a suitcase, getting out of bed or up from a chair — are usually followed by bad back pain that lasts for several months. Fractures are often the starting point of loss of confidence, loss of mobility and a general decline unless everybody works hard to prevent it.

What causes osteoporosis?

A cynical answer would be 'old age'! But to leave it at that shows no real appreciation of what we know, as well as with what we do not know, about the subject. Admittedly, osteoporosis is a multifactorial disease as far as causation is concerned. But this should mean that we must seek out the *causes* that we *can* identify and do our best to eliminate them by treatment whenever possible. Of course we cannot help being black or white, or women or men. But because we know that as far as women are concerned the menopause brings with it a period of special risk, then this risk factor can be tackled by the woman's doctor in terms of the treatment system known as hormone replacement therapy. We know also that bed rest and inactivity accelerate the whole process of bone absorption. Clearly, prolonged rest, particularly bed rest in the middle aged and elderly, is not helpful in combatting the disease. Active and prolonged physiotherapy helps to tip the therapeutic scales in the right direction.

Diet as medicine in osteoporosis

For years now physicians have paid lip service to the principle of the dietary management of osteoporosis and in a Consensus Conference on osteoporosis published in the *Journal of the American Medical Association* in 1980, an up-to-date but traditional view advocated increasing calcium intake to combat the disease. It pointed out that the average dietary intake of calcium in the United States was between 450mg and 550mg (UK figures are roughly similar). The National Research Council's RDA is 800mg. However, when it is remembered that RDAs are designed to meet the needs of only 95 per cent of the population, it would certainly seem sensible for everyone to increase their calcium intake to perhaps 1000 to 1500mg a day, in the interests of maintaining a positive balance of calcium all the time.

Medical/nutritional experts today are advising the population of the western world generally to reduce their fat intake. If people follow this advice, they will further reduce their calcium intake. The major sources of calcium in the diet are milk and dairy produce. Those who are following advice to eat with their cardiovascular health, or even their shape, in mind will often logically exclude milk from their diet as far as possible. But it is possible to minimize the low fat = low calcium equation by replacing ordinary milk with skim or low-fat milk. Dietary supplementation with elemental calcium poses a potential problem due to the possible encouragement of kidney stone production, particularly in those with a predisposition to or history of stone formation.

Normal levels of vitamin D are required for optimal calcium absorption too. Most diets contain plenty of vitamin D these days but people who do not receive adequate exposure to sunlight (those confined to home, and dark-skinned folk) may need supplemental vitamin D in daily doses of 600-800 units to help facilitate calcium absorption — but not more.

The contribution magnesium makes

Very slowly it is being realized that magnesium can make an important contribution to osteoporosis prophylaxis. Relatively little attention has been paid to the importance of magnesium in bone metabolism, other than the fact that it clearly influences the activity of the parathyroid glands. These tiny structures in the neck are very closely linked to bone mineralization. Experimental magnesium deficiency studies carried out on animals show how magnesium lack causes abnormalities in skeletal structure and enzyme and mineralization problems that resemble several common bone diseases including osteoporosis.

Earlier in this book it was noted that one of the reasons that doctors, generally speaking, were not over-enthusiastic about considering magnesium deficiency as a cause for many of the ills our flesh is heir to, is because blood tests for magnesium are usually so constant. But blood magnesium does not go up and down in magnesium deficiency like, for instance, iron in the blood does when the body becomes anaemic. This is because, it will be remembered, plasma levels of magnesium are maintained within very narrow limits, even in the face of insufficiency of magnesium intake or excessive losses. But still the body's magnesium is slowly mobilized from tissue stores in cases of magnesium imbalance.

The major depot of stored magnesium is in our bones. They contain two-thirds of our total body magnesium.

In young animals this bone-bound magnesium is relatively labile and easily mobilized. No doubt the bones in young people play a large part in keeping magnesium levels in the body stable. But as age increases, and especially if low-magnesium intakes persist, and if absorption is poor or dietary factors exist that compete for magnesium, then as tissue magnesium falls disturbances of bone modelling occur.

One of the reasons why high calcium intake is not all that effective in combatting osteoporosis is that high calcium intakes compete with magnesium for intestinal absorption. Excess of phosphate in the diet also decreases vital bone magnesium. We do not know the precise physiology of bone mineralization, or understand exactly what triggers the mechanism of osteoporosis, but Mildred Seelig, who has made an extensive study of the subject, writes in her fascinating book on the subject, *Magnesium Deficiency in the Pathogenesis of Disease — the early roots of cardiovascular skeletal or renal abnormalities*, that she believes that magnesium is heavily implicated:

> Bone wasting diseases that are resistant to physiological doses of vitamin D, calcium or phosphate and that have been treated with pharmacological doses of each or combinations of mineralizing agents (without positive results) are likely to be associated with magnesium deficiency.

Hidden until it strikes

One of the most miserable things about the osteoporosis story is that, as happened in the case of Ethel Green, quite often you don't know that osteoporosis is with you until it strikes and you fall down with a fracture. There is no quick and easy diagnostic test. Even X-rays that normally tell us so much about bone will only clinch a diagnosis when there is an advanced progress of the disease with a loss of 30-50 per cent of skeletal mass! This being so, earlier diagnosis must rest on a high level of suspicion — much higher than is prevalent generally today by patients and their physicians. Clearly, special risk groups should be very carefully monitored and energetically treated (for example, women at the menopause, and those who have had hysterectomy operations or who have suffered from premature menopause). It has been proved without doubt that the female sex hormone oestrogen allows women to retain more of their magnesium intake than young men can, but this facility disappears at the menopause. The reason that there is a later peak in osteoporosis problems in men than in women seems to be due to the fact that, hormonally at least men remain 'sexual' for much longer than women do, producing testosterone, the male sex hormone, well into old age. Both oestrogen and testosterone share several structural features chemically that are anti-osteoporotic factors.

In the past the common clinical symptoms of osteoporosis were ignored by doctors and patients alike. The general shrinking of the skeleton, the 'buffalo' hump at the base of the neck, the aches and pains in hands and arms brought about by compression effects on nerves as they leave the spinal cord for the arms, together with the observation that 'Grandma is growing downwards' are too often ignored. They tend to be shrugged off as part of the decline of advancing years. This was understandable perhaps when, generally speaking, specific treatment for osteoporosis was unsatisfactory, apart from hormone replacement therapy in menopausal women. But today, when it has been proved that magnesium supplements given to patients known to be magnesium deficient improve their calcium retention and that patients with osteoporosis retain *more* calcium when it is supplemented with magnesium, the time seems to have come when the magic of magnesium should be applied to the prevention of a nasty crippling disease.

CHAPTER 9

Shopping for Magnesium

The consensus of opinion seems to be that at best few of us get quite enough magnesium to keep us fit and well. To some extent the chance of geography exerts a powerful influence in this context. As we have seen, the beneficial effects of living in a hard-water area are particularly dramatic in this context — always remembering that to get the health bonuses from hard water you must drink it raw! Boiling, as any housewife who lives in a hard-water area will confirm, removes a lot of the 'good stuff' from nice hard tap water and leaves it in the kettle as scale!

But where you live or rather, where your food comes from — also influences whether or not your foodstuffs contain all the magnesium that Nature intended. Intensively farmed soils soon become exhausted of their mineral content unless they are repeatedly and *organically* fertilized. Many cheap and commonly used inorganic fertilizers replenish the necessary potassium, nitrogen and phosphates plants need for rapid commercial growth, but because they are lacking in magnesium they produce magnesium-deficient produce.

Looking for magnesium

The main availability of magnesium in food is shown on page 32, but this could be a snare and a delusion. To start with, lists of mineral-rich foods seldom if ever take any account of the important geography of crop production. Secondly, losses incurred during processing and cooking of food are usually ignored too. But in the case of magnesium there is also another factor to be considered.

A long-term study on magnesium balance carried out at a Veterans' Hospital Metabolic Unit in the USA produced some very interesting and pertinent facts about magnesium *absorption*. To start with, just increasing the magnesium intake in the diet did not consistently increase the amount of magnesium retained in the body because the amount of other minerals in the diet influenced magnesium absorption profoundly. Patients eating a diet containing 1400mg of calcium per day (more than the RDA of 800mg) remained in negative magnesium balance even if they were given a magnesium supplement of 500mg per day. Not until their *calcium intake* was raised to above 2000mg per day were these subjects eased into a comfortable positive magnesium balance.

But an even more interesting state of affairs was revealed when the scientists involved in this metabolic study turned their attention to the *phosphorus intake* of their patients, now so carefully manipulated into positive and 'safe' magnesium balance. When this happy state of affairs obtained there was a little under 1000mg of phosphorus in the daily diet. If this was subsequently increased to 1500mg the positive balance of magnesium (of 29mg per day) previously enjoyed was converted into a negative one (of 19mg) simply due to the effect of phosphorus on magnesium uptake. In other words, when seeking foods for their magnesium content, attention also needs to be paid to the phosphorus content of those foods (see the table below showing high phosphorus content foods). At the same time a calcium intake in excess of 2000mg per day seems necessary.

High Phosphorus Content Foods
(mg per 100g edible portion)

Food	*mg*
Baker's yeast	1300
Wheat bran	1250
Milk (dried skimmed)	950
(dried whole)	740
Cocoa powder	660
Brazil nuts	600
Sesame seeds	600
Soyabeans	550
Soya flour	550
Walnuts	510
Pistachio nuts	500
Cheese, hard	500

Food	*mg*
Almonds	500
Cod's roe	500
Wheat flour (self-raising)	480
Sardines (canned)	425
Cashew nuts (roasted)	400

Also to be borne in mind in this context is the calcium to phosphorus ratio. This first started to interest doctors when they realized that increased phosphorus in food reduces the calcium balance and so this had to be considered when calcium intake was being studied. Now that the significance of new knowledge on the effect of phosphorus on magnesium uptake from the diet has come to light, the C/P ratio of foodstuffs as it is now being referred to, has become very interesting. In fact, it has been suggested by research workers from the University of California at Los Angeles medical school that an overall C/P ratio of between 1:1 to 1:1.5 is reasonable for adults to provide good calcium absorption.

Magnesium in the shopping basket

So, from the practical point of view, shopping with magnesium in mind is not such a straightforward matter as it seemed when we first looked at the magnesium content of our food. Many of the magnesium-rich foods we want to remove from the supermarket shelves and place in our shopping basket, such as nuts and wheat products (which have more than 100mg magnesium per 100g) now look very much less attractive because they add considerably to our phosphorus intake.

Seafood looks pretty good on this score provided we exclude scallops and crab. Many of the really rich magnesium foods (cocoa, chocolate and nuts) tend to get excluded automatically from the diets of slimmers because of their high calorific content and so do not contribute much to the overall magnesium intake. Most green vegetables contain moderate amounts of magnesium when harvested. But enormous differences between 'on the shelf' magnesium and 'on the plate' magnesium occur due to the very variable treatments that vegetables get in the kitchen. We all know what boiling does for hard water. In all probability it does the same for vegetables too.

Many of our favourite staple foods like meat, offal, white fish such as cod and halibut, together with chicken and turkey (now becoming increasingly popular because they have a lower fat content than red meat) are relatively poor in magnesium content (under 25mg per 100g). Fruit is generally speaking, a popular 'health food' because it is low in fat and calories and high in certain vitamins. But it too scores poorly as far as magnesium is concerned.

Mildred Seelig, who probably has spent more time and energy weighing up the content of magnesium in food than anyone else, supports the contention that the magnesium supplied by the American diet, and probably by that of most industrialized countries, *particularly those populated by Europeans or by those who have comparable eating habits,* is likely not to be optimal. Intakes which often are at best marginal she feels can be frankly deficient where they are concomitant with high intakes of nutrients that increase magnesium requirements.

Dr Seelig seems puzzled as to why some of us develop illnesses related to low tissue magnesium (such as cardiovascular diseases, kidney stones and pre-menstrual tension), while other folk eating the same sort of diet remain fit. She often suggests that family or group eating habits may be implicated. It is far far more likely that the basic cause lies in the relative efficiency of our individual absorption organs. This subject has been more fully discussed in my book on Selenium referred to earlier.

All in all, the complexities of good shopping for magnesium in the supermarket or elsewhere are such that many sensible folk are tending to make sure their tissues are in safe magnesium balance by the taking of a reputable magnesium supplement daily.

CHAPTER 10

Pregnancy, Magnesium and Your Children

Everybody realizes that pregnancy is a time when the whole woman's physiological emphasis is centred around the formation of new tissues. For example, a larger womb with its ever-growing placenta, bigger breasts, more blood and of course the development of the little 'parasitic' passenger, her eagerly awaited baby. The mother needs more food during pregnancy, of course. She is eating for two without doubt. But she must be careful not to assume that she needs to eat twice as much as she did! If she does this, she is likely to put on excess weight during pregnancy and this is an unhealthy thing to do. Being overweight in pregnancy is as hazardous medically, for mother and child, as being under-nourished.

Only comparatively recently, however, has the mineral nutrition of women during pregnancy been a subject of concern for doctors, nutritionists and at the ante-natal clinic. Before that ideas, particularly with reference to the ideal magnesium status during pregnancy, were based on 1914 studies! Now the United States and Canada Food and Nutrition Boards have put their heads together and recommended a daily intake of 450mg of magnesium per day for pregnant women.

This relatively small recommended increase over the RDA norm (300mg) has come about as the results of various metabolic studies on magnesium balance in pregnant women are becoming more widely known. Many of these have demonstrated quite striking deficiencies of magnesium occurring during pregnancy. Strangely perhaps, a surprising piece of 'hard evidence' on this score came about as a result of a medical scare about the taking of milk of magnesia as a laxative during pregnancy.

Some physicians were worried in case it interfered with calcium retention by the mother and infant during pregnancy. This could predispose towards rickets, they thought. Happily, it was demonstrated that milk of magnesia produced no interference at all in calcium retention but other interesting information did come to light as a result of investigations. Milk of magnesia is rich in magnesium, of course. But the studies on pregnant women, even taking relatively high 'supplements' of magnesium (as a pregnancy laxative), showed that as pregnancy progressed there was a tendency for larger and larger negative balances of magnesium to occur. This demonstrated for the very first time just how much extra magnesium appears to be needed during pregnancy.

Incidentally, other metabolic studies have demonstrated that the addition of vitamin D to the mother's diet (in the correct amount) tends to improve magnesium and calcium balance in the pregnant women. Probably as a result of this nutritional news way back in the 1930s, cod liver oil started to be widely recommended to pregnant women. This was given on the grounds of providing mother and baby with sound bones and teeth through adequate calcium balance rather than being anything at all to do with maintaining a healthy magnesium balance, for knowledge of the magic of magnesium was in its infancy at this time.

The parasitic foetus

Much has been written about the irresistible demands on the mother that are made by the baby in her womb. Even in the starvation camps of Nazi Germany during the Holocaust, babies were born which although feeble and under-nourished, in no way matched the wasted state of their poor mothers. When the dietary intake of magnesium is not sufficient to meet the demands of the developing baby, maternal stores of the mineral are mobilized and a magnesium deficiency develops in the mother.

One of the most baffling complications of pregnancy is the disease known as eclampsia or toxaemia of pregnancy. This is a potential killer of women and their babies. But modern medical management, which involves strict bed rest, the lowering of blood pressure, a strict control of body weight coupled with an admirable system of 'early warning' monitoring of very early toxaemia symptoms (such as urinary protein excretion, raised blood pressure, and excessive pregnancy weight gain) has reduced this potential 'terror' of pregnancy to the status of a much milder danger.

Incredibly perhaps we still do not know for sure what *causes* toxaemia of pregnancy for, despite its name, no toxin, or poison, has ever been discovered that is associated with it. Significantly, however, magnesium may be involved in this mystery. Way back in the early days of this century, when doctors were fighting toxaemia in their patients often in the full-blown, so called *eclamptic* form in which convulsions so often herald the death of a toxaemia victim — they started to give injections of magnesium. Often as if by magic the fits would stop. Only comparatively recently it has been shown that magnesium levels in toxaemic women are lower than their non-toxaemic sisters during pregnancy and the basic cause of toxaemia may well be a degree of magnesium lack.

Is morning sickness to blame for more than we think?

An extensive review by Dr Mildred Seelig has shown that magnesium intake in the diet of pregnant women generally is likely to be suboptimal. That it might be low enough to make a contribution to several early or late abnormalities of pregnancy was suggested by more recent work by Drs Johnson and Philipps in the United States, while Egyptian researchers who have studied magnesium levels in pregnant women have suggested that morning sickness, the very common accompaniment of pregnancy, might contribute clinically to magnesium deficiency. They also noted that there was a direct relationship between low magnesium intake and low birth-weight infants who had a poor survival outlook.

There is no doubt that morning sickness, because of the withdrawal of most remedies for the condition due to the 'Debendox scare', now receives little in the way of medical treatment. This makes the likelihood of magnesium depletion early in pregnancy much stronger than it was a few years ago. There is also no doubt that a magnesium-depleted mother is highly likely to produce a magnesium-depleted baby.

The exact part that magnesium depletion plays in diseases of infancy is really only just being seriously considered, but at an International Symposium on magnesium recently, a Scottish group of workers headed by Dr J. O. Forfar suggested that disturbed magnesium metabolism may well play a significant role in convulsions occurring in otherwise normal infants. What is more, they have backed up this hypothesis by presenting evidence that both plasma and cerebrospinal fluid levels of magnesium and calcium are lower in convulsing infants than in normal infants.

Babies that are prone to suffer so-called spontaneous convulsions are often described as being jittery babies. They startle easily, are easily upset and so on. The group in Scotland who have carried out most of the research work in this field have coined the term 'convulsions due primarily to mineral deficiency' to distinguish the diagnosis in such cases from convulsions that are liable to be due to high temperature problems or even epilepsy. They based their conclusions on the investigation of seventy-five new-born babies with fits seen over a period of two years. In such babies 92 per cent had subnormal calcium levels and 52 per cent had subnormal magnesium levels.

The part that patterns of infant feeding contribute to these 'mineral' fits is highly interesting. Generally speaking, infants fed on evaporated cows' milk formulas have low magnesium and high phosphorus levels in their blood. This is of course entirely to be expected. The mineral content generally of human milk is considerably less than that of cows' milk because the calf grows very much more quickly than a human baby does. Cow's milk is particularly rich in phosphorus and this excessive phosphorus contributes to the abnormalities of serum levels in both calcium and magnesium in bottle-fed babies.

The first baby whose convulsions were found to be related to low levels of magnesium in his tissues, and low blood levels of calcium in his blood seemed to be a chronic malabsorption victim for, although his fits responded to magnesium, it needed five times the normal intake of magnesium to keep him convulsion free. If supplemental magnesium was withheld, decreases in his magnesium levels immediately started to occur.

Gastroenteritis hazards

The management of infantile gastroenteritis has improved enormously over the last few years due to the introduction of electrolyte replacement therapy. Prior to this, gastroenteritis was one of the commonest causes of death in infancy. One of the constant features of infantile gastroenteritis to be noted was that it was almost always bottle-fed babies who contracted the disease, while breast-fed children were virtually immune to it. Making assumptions in clinical matters is a dangerous business and the assumption so frequently made in this case was that the reason the formula feeds triggered off gastroenteritis in babies was that such feeds were contaminated (carelessly) while they were being made up. Obviously such a state of affairs did occur occasionally. But in many cases the rapid collapse of babies who developed diarrhoea was due to factors other than germs and microbes. Today we know that electrolyte depletion and subsequent dehydration is usually what really kills babies with serious gastroenteritis and that, provided they are supplied with plenty of fluid by mouth containing a good physiological 'mix' of sodium/potassium and glucose, they can rapidly absorb the fluids they so vitally need when they suffer from diarrhoea and vomiting.

However, it has also been demonstrated that young babies are particularly vulnerable to magnesium depletion. The composition of formula feeds has the tendency to put them at hazard nutritionally because of the relatively high phosphorus level of such feeds. Then if they suddenly find themselves in a gastroenteritis situation they are in double jeopardy if their stores of magnesium are low.

Most babies recover from their gastroenteritis rapidly on electrolyte replacement therapy. A few, however, still have to be admitted to hospital to be treated with intravenous fluids to replace their fluid loss. Some such babies do not start to recover until magnesium is added to the intravenous fluids they are receiving.

Sudden Infant Death Syndrome (SIDS)

In Chapter 1 we met a small group of people who were helped in one way or another by magnesium. One member of this group was rather different from the rest. Her name was Adrienne and she had experienced something that is dreaded above all others by everyone who is involved in raising a family. She had put a healthy baby to bed one night and a few hours later had found the self-same baby dead in its cot — a cot death or, as it is now being referred to more commonly, a case of Sudden Infant Death Syndrome (SIDS).

After all the agony of coping with her dreadful shock and heartbreak, Adrienne really only wanted to do one thing — have another baby and raise it safely and successfully. To start with, she was terrified to begin another pregnancy because nobody could really convince her that there was no chance of SIDS happening again. 'Even lightning can strike the same place twice', as one doctor put it. And then she met a woman doctor who, while pointing out that although the whole question of SIDS is still a mystery, it does look as though magnesium might have a part to play, and might even be a SIDS prophylactic to some extent.

This was the substance of the advice which convinced Adrienne that she would 'risk getting pregnant' again. The classic typical case of SIDS reflects everything that had happened in Adrienne's tragedy. The SIDS baby seems healthy and is growing well. There is nothing more seriously medically amiss than a slight cold-like respiratory infection and a mild feeding difficulty for a day or two before the baby is found dead in its cot. Usually these tragedies occur in winter or spring. In Adrienne's case it was 1st March. Most often the mother is young, although she may already have a baby or two. Adrienne was 21. Recent studies suggest that the incidence of SIDS is more common than we think — there is perhaps one SIDS death to every 400-500 births.

In the past, parental neglect or incompetence have been suggested causes of SIDS, but really there is no convincing evidence to support this theory. One theory was that SIDS babies somehow turn over in their cots and asphyxiate themselves. This has been similarly questioned. The major theories popular today are that SIDS is caused by a sudden overwhelming virus infection, or perhaps that the SIDS baby has a poorly developed immunity system, or even that its autonomic nervous system is somehow defective. All these theories suffer from one objection or another, the most powerful being that SIDS does not seem to occur in a recurrently ailing baby, which would be expected if immunological or nervous system defects were involved. It happens to perfectly fit babies. In the 1970s, doctors started to consider that cardiac (heart) abnormalities might be involved in SIDS babies. Then a fortuitous happening drew Dr J Salk and some colleagues to make a report that was published in the prestigious *New England Journal of Medicine* in 1974. This focused medical attention more firmly on the heart in SIDS.

It happened like this. A group of twenty-four healthy infants were being studied with reference to the physiological reactions of their heart rate. Suddenly one of the children died of SIDS. Quite by chance it had previously been noted that this baby's heart rate was significantly deviant from that of the other twenty-three infants studied, although no sign of heart disease was present.

Then there was a Dr J Caddell who was one of a small group of children's specialists interesting themselves in the possibility of magnesium deprivation being implicated in SIDS. Dr Cadell had in fact published an account of her studies in the *Lancet*. She had pointed out that infants with low birth weights also had low magnesium stores and were more vulnerable to SIDS than other babies. She also pointed out that the dread disease is now more common in formula-fed, rapidly-growing infants. Such children have high calcium and high phosphorus levels in their tiny bodies while their magnesium status is poor.

Several other medical and nutritional authorities have augmented this new outlook on SIDS, pointing out an association between SIDS and mothers who are themselves low in their magnesium status and who have suffered from pre-eclampsia, or toxaemia of pregnancy or who have had babies rather close together — all factors associated with negative magnesium balance.

As we have seen elsewhere in this book, magnesium deficiency is very often associated with minor cardiac abnormalities. Many physicians have related poor magnesium status to heart attack, particularly the 'coronary thrombosis without a thrombus' syndrome that we met in Chapter 4.

Clearly we are not out of the scientific wood as far as SIDS is concerned and there is need for further research in this matter. Unfortunately, magnesium determinations are almost never reported in mothers or siblings when infants die of SIDS. Nor, for that matter, are in babies who die of heart failure or even congenital heart disease.

To some extent this regrettable state of affairs is the result of medical apathy about investigating the mineral status of tiny babies. But, to be fair to the doctors, the obtaining of useful information about magnesium status is, as we have mentioned before, a complex business and not the sort of 'push-button' medical test that doctors are happy to request routinely. At the moment determination of magnesium status means assessing a child's response to a magnesium load. This in turn means an intravenous injection, and many doctors would consider such a test to be too 'invasive' to be part of even a research programme.

Adrienne, as we saw in Chapter 1, gained considerable confidence by taking, on her doctor's advice, supplementary magnesium all through her pregnancy. She then stood an excellent chance of going into labour in positive magnesium balance and having a baby born with excellent magnesium stores in its tiny body. She was also determined to breast-feed her baby and to continue taking a magnesium supplement all through her lactating period. She realized that nothing can provide a guarantee against SIDS, but at least she was doing everything possible to prevent such a tragedy happening again.

CHAPTER 11

Postscript: Pills and Pregnancy

These days women are quite rightly urged to take nothing in the way of self-medication if they either know or suspect that they are pregnant, and to inform any physician who may be attending them of their situation. This particular state of affairs has come about as a result of what is still looked upon as the thalidomide scandal.

It is perhaps unfortunate that the word 'scandal' was applied to this context. The whole thalidomide affair was both unfortunate and tragic in very many ways, but to continue to refer to it as something scandalous has been counter-productive as far as health is concerned for it has produced an aura of fear about medication that is well in excess of common sense.

Recently, this has extended itself into the field of nutrition as well as medication, with unfortunate results. For example, women have been known to enter into labour at the end of pregnancy with dangerously low levels of vital haemaglobin because 'thalidomide fear has made them throw away all the iron pills that had been prescribed for them during pregnancy. A straight look at the tragedy of thalidomide without fear or prejudice may help to clear the air on this much misunderstood matter.

The thalidomide babies

Most people know that as a result of thalidomide the babies involved developed various abnormalities, usually of the limbs. Doctors call this the *teratogenic* effect of thalidomide and *tera* is Greek for monster. The poor little thalidomide children were hardly monsters, but they were malformed and the study of malformed individuals generally is graced by the name of teratology in medical parlance.

Since the beginning of time, people have felt a mixture of awe, wonderment and fear at the birth of any abnormal infant. Ancient peoples proposed many explanations for the birth of malformed individuals and blamed the moon, the stars and even divine intervention to explain away something that we now know happens spontaneously to 3-6 per cent of all new-born children.

In the past, a less than perfect baby was often thought to be due to the breaking of some religious or social taboo. For instance, the ancient Hebrews and Romans believed that such babies were born as a result of sexual intercourse having occurred during a menstrual period — a belief that persisted into the seventeenth century in Europe.

With the Renaissance came yet another explanation of *teras* — the theory of maternal impressions, based on the idea that a fright or sudden shock suffered by the mother during pregnancy produced a less than perfect child or even a monster. All these curious facts from antiquity are worth remembering because they all reflect the puzzlement, anxiety and guilt that is associated with the birth of a less than perfect or a deformed baby. Curiously, it seems to be a common trait in human nature that if you can blame your misfortune on somebody else or something else then somehow or other you feel rather better about it yourself.

The beginning of this century saw a greater understanding of the mechanics of genetics, and the dawn of modern embryology brought science to the subject for the very first time. Way back in 1905, one type of abnormality called brachydactyly (short fingers) was shown to follow the same principle which Mendel had described as producing short and tall plants, and so terms like 'recessive genes', 'mutations', 'sex-linked' and so on extended the language of medical science. Quite recently, geneticist James Wilson estimated that between 20-25 per cent of all malformations are caused by such natural genetic defects.

This leaves about three-quarters of serious abnormalities happening with no known cause. By the end of the first quarter of our century however, the concept that certain environmental agents (now christened teratogens) could cause malformations stood on pretty firm ground. Curiously perhaps, some of the earliest work in this field concentrated on *deficiencies* of nutrients being teratogenic. For instance, vitamin A deficiency in sows resulted in all sorts of malformations in the piglets of those sows. Then in 1941, Dr Norman Gregg demonstrated that a virus infection (German measles) could act as a potent teratogen in mankind.

By and large, however, the scientific world did not accept that drugs could act as teratogens until 1960, when the discovery was made that a synthetic drug called thalidomide, which was being taken by pregnant women in Germany, the UK, Australia and Japan, was a powerful teratogen.

Like all medicines, thalidomide was developed as a drug with the very best intentions. Marketed under a variety of trade names in the late 1950s, it proved to be a remarkable and safe sedative (transquillizer) and 'hypnotic', by which doctors mean a sleep-inducing pill.

At this time, the most popularly prescribed sedatives and sleeping tablets were in the class of drugs known as barbiturates. They worked well enough but became associated with two lethal, or potentially lethal, complications — suicide and accidental overdosage. Barbiturates kill because of the ease with which they knock out (depress) the brain's respiratory centre, a tiny area in the brain that automatically controls our breathing mechanism and rate.

An overdose of sleeping tablets (which could be as few as half a dozen or so pills, depending on strength) pushed countless numbers of suicidally inclined people out of this life.

But not all sleeping-pill deaths were suicidal and another common cause of death during the barbiturate era was accidental overdosage. It came about like this. Barbiturates not only cause sleep and heavy tranquillization, they also cause amnesia (loss of memory). What usually happened was that a habitual sleeping-pill taker would swallow the usual dose of sleeping pills, then half an hour later he would realize that the pills had not worked and repeat the dose (barbiturate habituees gradually developed a high degree of tolerance to their 'sleepers' and tended to take increased doses or to repeat doses as described). In some cases in a state of drug-induced amnaesia, people would repeat the dose several times. A fatal outcome quite often seemed to occur when a 'double dose' — say at 11 p.m. — was followed by another similar dose at, say, 11.30 p.m. By midnight, the person was asleep but by 1 a.m. his respiratory system had failed and he was dead — by accident.

Thalidomide was an excellent sleeping pill and yet even with huge doses it did not 'knock out' the breathing centres in the brain. Understandably, its prescription rapidly became popular with doctors, who themselves often felt guilty when their patients 'overdosed'. The traditional historians of thalidomide often fail to note the next development. It was found while the drug was being developed that thalidomide, as well as being a safe sedative, was also an excellent anti-emetic. In another words, it stopped you being sick. Now, of course, pregnant women are often sick and in the 1950s were often prescribed antihistamines to make life more comfortable for them in early pregnancy. But accidental overdoses sometimes caused problems here too. Once the new, apparently safe drug, thalidomide, became available, it was tempting to use this as a new and safer pregnancy sickness remedy. And so two classes of expectant mothers — the insomniacs and those who experienced bad pregnancy (morning) sickness — were unknowingly exposed to a new risk, thalidomide teratogenicity, for what was, considering the state of pharmaceutical knowledge at the time, the best possible motives.

Then in 1961 two doctors, one in West Germany and the other in Australia, reported in the scientific literature and to thalidomide's manufacturers the fact that the drug thalidomide was producing birth defects in mothers who had taken it during their pregnancy. The drug was rapidly withdrawn. But by now the thalidomide damage was done.

Even today the age-old guilt and fear reactions that we noted earlier with reference to the birth of malformed children are alive and well. Quite soon, too, the bad news about thalidomide became 'good news' for the 'investigative' type of journalist who immediately saw a heaven-sent opportunity to make claims that the real villain of the whole thalidomide affair was the Pharmaceutical Industry who had launched an improperly tested drug on an unsuspecting public. An impartial appraisal of the facts of the matter, however, must tell a very different story.

Knowledge of chemical teratogens was in its infancy prior to 1961. Nevertheless, thalidomide, like all drugs, was tested in pregnant laboratory animals for teratogenicity using the usual laboratory animals kept for such tests. Thalidomide passed all these tests with flying colours. Why all the thalidomide babies, therefore? The answer was found to reside in what was in 1961 an entirely new concept. This became known as *teratogenic species specificity*. Humans and non-human primates are sensitive (teratogenically) to thalidomide. Rabbits are far less sensitive. Rats and mice, the almost universal laboratory test animals, are essentially *insensitive* even to *large doses* of thalidomide. And so the teratogenic effects did not show up in routine testing.

In many ways it seems that thalidomide had to happen. Or, to put it another way, if the tragedy had not happened with thalidomide it would have happened at some later date with another new drug and perhaps with even more devastating results. Thalidomide did not only produce a lot of deformed children, it also launched an enormous explosion of research into the whole mechanism of teratogenicity — research that has already paid fine dividends as far as health is concerned.

Vitamins, minerals and health hazards

As stated previously, phobic and irrational fear of 'another thalidomide' still haunts our culture. It has already caused one very effective anti-nausea remedy (Debendox) to be removed from world-wide markets without any real hard scientific evidence being mounted against it. The same sort of phobic reaction has also made many folk frightened of taking extra vitamins, and put pregnant women off taking minerals like iron, often to the detriment of their health and well-being during pregnancy.

Incredibly, all the evidence as far as vitamins and minerals are concerned has been on the credit side. For instance, a large survey into the effects of vitamins and the incidence of neural tube defects like spina bifida showed that mothers who took multivitamin supplements during pregnancy tended to produce fewer babies suffering from this devastating type of deformity than women who took no such multivitamins. Critics of this study are attempting to mount a large and far-reaching study to confirm or deny the useful prophylactic effect of vitamins during pregnancy. But serious ethical problems have recently been quite rightly raised on the grounds of whether it is justifiable to condemn the 'control' women in the study (who would take *no* extra vitamins during their pregnancy) to a *higher* risk of having a malformed baby than if they had taken a daily vitamin supplement.

At the beginning of this chapter, I stated what must be a worthy and sensible rule for all women to follow. If they are pregnant or trying to become pregnant they should consult their doctors before swallowing any medicines, either prescribed or obtained over the counter at a pharmacy or chemist. Does the same stringency apply to health foods and nutritional products? Nobody has categorically stated as much, but an increasing consensus of medical opinion inclines itself to the view that such products only do good and are utterly and completely safe during pregnancy — as indeed food itself is! But if you are still worried, then the old dictum of 'best ask your doctor' is still an excellent one — and as far as the UK is concerned it is completely free as well!

Index

(bold entries refer to diagrams)